MATERIALS AND COATINGS
TO RESIST
HIGH TEMPERATURE CORROSION

Papers presented in May, 1977, at a two-day meeting in Düsseldorf, West Germany, organized by the Verein Deutscher Eisenhüttenleute (German Steelmaking Association) on behalf of the European Federation of Corrosion's Working Group on Corrosion by Hot Gases and Combustion Products

MATERIALS AND COATINGS TO RESIST HIGH TEMPERATURE CORROSION

Edited by

D. R. HOLMES

*Materials Division, Central Electricity Research Laboratories,
Leatherhead, Surrey, UK*

and

A. RAHMEL

*High Temperature Corrosion Group, DECHEMA Institute,
Frankfurt/Main, West Germany*

APPLIED SCIENCE PUBLISHERS LTD
LONDON

APPLIED SCIENCE PUBLISHERS LTD
RIPPLE ROAD, BARKING, ESSEX, ENGLAND

British Library Cataloguing in Publication Data

Materials and coatings to resist high temperature corrosion.
  1. Corrosion resistant materials—Congresses
  2. Metals at high temperatures—Congresses
  3. Protective coatings—Congresses
  I. Holmes, D   II. Rahmel, A
  620.1′6′23   TA418.75

ISBN 0-85334-784-0

WITH 26 TABLES AND 265 ILLUSTRATIONS

ⓒ APPLIED SCIENCE PUBLISHERS LTD 1978

Printed in Great Britain by Galliard (Printers) Ltd, Great Yarmouth

Preface

This book contains papers presented at an international conference of the European Federation of Corrosion's Working Group on Corrosion by Hot Gases and Combustion Products, which took place on 16 and 17 May, 1977 in Düsseldorf and was organized by the Verein Deutscher Eisenhüttenleute (German Steelmaking Association). The Editors would like to thank the conference organizers on behalf of the authors and all the participants for their successful work and for the support they gave to the Working Group. Simultaneous translation of the presentations and discussions helped to minimize language barriers between the participants from 12 different countries.

This was the eighth conference of the Working Group since its formation in 1966. Earlier conferences covered the themes 'Corrosion of Metals in Molten Salts' (Düsseldorf, 1968), 'The Role of Chlorine in the Corrosion Caused by Deposits' (London, 1969), 'Influence of Trace Elements on the Oxidation of Metals and Alloys' (Düsseldorf, 1970), 'Mechanical Properties and Adhesion of Oxide Films and their Influence on the Oxidation of Metals' (Düsseldorf, 1971), 'Deposition and Corrosion in Gas Turbines', (London, 1972), 'Techniques to Minimize High Temperature Corrosion by Additives, Fuel Treatment and Coatings' (Copenhagen, 1974), and 'High Temperature Corrosion in Gases with Two Aggressive Components' (Düsseldorf, 1975). Generally the proceedings of these conferences are not published because it is a policy of the Working Group that preliminary results not yet ready for publication should be discussed in an informal atmosphere. Therefore only the papers of the conferences in 1971 (Verlag Chemie, Weinheim, FRG) and in 1972 (Applied Science Publishers) have been published in book form. For the 1977 conference the authors wished them to be published in this book; they have been supplemented by the paper by Dr Elliott and Mr Taylor which could not be presented at the conference.

The conference covered both bulk materials and coatings for high temperature service. The coatings can be considered as special high temperature materials, because, besides exceptional oxidation and corrosion resistance, they also need certain minimum levels of creep

v

resistance and other mechanical and physical properties. For convenience we have arranged the papers in the book in the same way as they were presented at the conference, Chapters 1 to 11 dealing with high temperature materials and Chapters 12 to 23 with coatings. Important parts of the discussions which followed the papers have been included.

The book stems from a happy co-operation between the conference organizers, the authors of the papers, the publishers, and ourselves the Editors. We should like to thank our colleagues for making its production possible and the conference participants for providing their discussion contributions. We hope that the readers of the book will find its contents worthwhile and useful in their work.

D. R. HOLMES
Materials Division,
Central Electricity Research Laboratories,
Leatherhead, Surrey, UK

and

A. RAHMEL
High Temperature Corrosion Group,
DECHEMA Institute, Frankfurt/Main,
West Germany

Contents

Preface v

List of Contributors xi

1. Influence of the Composition of Fe–Cr–Al Ferritic Steels on the
   Corrosion Processes and Rates in Sulfur at High Temperature   1
   J. P. LARPIN, M. LAMBERTIN and J. C. COLSON (*Université de
   Dijon, France*)

2. Comparative Studies of the High Temperature Oxidation of
   Four Austenitic CrNi Steels in Helium with Water Vapor and
   Hydrogen Additions 21
   S. LEISTIKOW, R. KRAFT and E. POTT (*Institut für Material- und
   Festkörperforschung, Karlsruhe, West Germany*)

3. Influence of Zirconium as an Alloying Addition on the
   Corrosion Behaviour of a Ni/15Cr Alloy in an Oxygen/Sulphur
   Dioxide Atmosphere at 850°C 39
   K. N. STRAFFORD and P. J. HUNT (*Newcastle upon Tyne
   Polytechnic, UK*)

4. Effect of Yttrium and Hafnium Additions on the Oxidation
   Behaviour of the System Co–Cr–Al   55
   D. P. WHITTLE, I. M. ALLAM and J. STRINGER (*University of
   Liverpool, UK*)

5. Effect of Grain Size and Y_2O_3 Additions on the Hot Corrosion
   Behaviour of IN-738 71
   P. HUBER (*Sulzer Brothers Ltd, Winterthur, Switzerland*) and
   G. H. GESSINGER (*Brown Boveri Research Centre, Baden,
   Switzerland*)

6. Effect of High Temperature Corrosion on the Creep Behaviour of Aero Gas Turbine Materials Under Service-Like Conditions 87
KH. G. SCHMITT-THOMAS, H. MEISEL and H.-J. DORN (*Technische Universität, München, West Germany*)

7. Comparative Studies of Corrosion Due to Hot Salts or Combustion Products of Kerosene Containing Impurities . 105
Y. BOURHIS and C. ST. JOHN (*l'Ecole Nationale Supérieure des Mines de Paris, France*) and R. MORBIOLI and S. FERRE (*SNECMA, France*)

8. Corrosion Above 700°C in Oil-fired Combustion Gases . 121
S. BROOKS, J. M. FERGUSON, D. B. MEADOWCROFT and C. G. STEVENS (*Central Electricity Research Laboratories, Leatherhead, UK*)

9. Corrosion–Erosion and Protection Techniques in Furnaces for the Incineration of Household Refuse 139
A. MOREAU (*Electricité de France, Paris, France*)

10. Corrosion Behaviour of Hot-pressed and Reaction-bonded Silicon Nitride in Condensed Phases 161
E. ERDOES and H. ALTORFER (*Sulzer Brothers Ltd, Winterthur, Switzerland*)

11. Application of the Acoustic Emission Technique to the Detection of Oxide Layer Cracking During the Oxidation Process 175
C. CODDET, G. BÉRANGER and J. F. CHRÉTIEN (*Université de Technologie de Compiègne, France*)

12. Profile of Requirements for Protective Coatings Against High Temperature Corrosion and Some Experience with their Behaviour in Service 185
W. BETZ (*Motoren- und Turbinen-Union, München, West Germany*)

13. Coating Requirements for Industrial Gas Turbines . . 199
P. C. FELIX (*Brown Boveri & Cie, Baden, Switzerland*)

14. Phase Stability of High Temperature Coatings on NiCr-base
    Alloys 213
    U. W. HILDEBRANDT, G. WAHL and A. R. NICOLL (*Brown
    Boveri & Cie AG, Heidelberg, West Germany*)

15. Ductile–Brittle Transition of High Temperature Coatings for
    Turbine Blades 233
    A. R. NICOLL, G. WAHL and U. W. HILDEBRANDT (*Brown
    Boveri & Cie AG, Heidelberg, West Germany*)

16. Diffusion and Precipitation Phenomena in Aluminized and
    Chromium-aluminized Iron- and Nickel-base Alloys . . 253
    E. FITZER and H.-J. MÄURER (*Universität Karlsruhe, West
    Germany*)

17. Influence of the Mode of Formation on the Oxidation and
    Corrosion Behaviour of NiAl-type Protective Coatings . 271
    R. PICHOIR (*ONERA, Chatillon, France*)

18. High Temperature Behaviour of Protective Coatings on Ni-
    base Superalloys 293
    P. C. MARTINENGO and C. CARUGHI (*FIAT (CRF), Orbassano,
    Italy*) and U. DUCATI and G. L. COCCIA (*Polytechnic School of
    Milan, Italy*)

19. Reaction-sintered Ni–Cr–Si Coatings on Nickel Alloys . 313
    E. FITZER, W. NOWAK and H.-J. MÄURER (*Universität Karlsruhe,
    West Germany*)

20. Preparation and Investigation of Layers Enriched in Silicon by
    Chemical Vapor Deposition 333
    G. WAHL and B. FÜRST (*Brown Boveri & Cie AG, Heidelberg,
    West Germany*)

21. Some Aspects of Silicon Coatings under Vanadic Attack . 353
    P. ELLIOTT and T. J. TAYLOR (*UMIST, Manchester, UK*)

22. Hot-corrosion Behavior of Chromium Diffusion Coatings . 369
    R. BAUER, H. W. GRÜNLING and K. SCHNEIDER (*Brown Boveri
    & Cie, Mannheim, West Germany*)

23. Preparation and Oxidation of Zirconium Silicide Coatings on
Zirconium 387
M. CAILLET, H. F. AYEDI, A. GALERIE and J. BESSON (*Institut
National Polytechnique de Grenoble, Saint Martin d'Hères,
France*)

Index 399

List of Contributors

I. M. ALLAM
Department of Metallurgy and Materials Science, University of Liverpool, P.O. Box 147, Liverpool L69 3BX, UK.

H. ALTORFER
Gebr. Sulzer AG, Metallkunde, Abt. 15, CH-8401, Winterthur, Switzerland.

H. F. AYEDI
Departement de Chimie, Faculté des Sciences et Techniques de Sfax, Tunisia.

R. BAUER
Brown Boveri & Cie AG, Postfach 351, D-6800 Mannheim 1, West Germany.

G. BÉRANGER
Département de Génie Mécanique, Université de Technologie de Compiègne, BP 233, F-60 206 Compiègne, France.

J. BESSON
Laboratoire d'Adsorption et Réaction de Gaz sur Solides, Equipe associée au CNRS No. 368, Ecole Nationale Supérieure d'Electrochimie et d'Electrométallurgie, BP 44, Institut National Polytechnique de Grenoble, 38401 Saint Martin d'Heres, France.

W. BETZ
Motoren- und Turbinen-Union, Postfach 50 06 40, D-8000 München 50, West Germany.

Y. BOURHIS
l'Ecole Nationale Supérieure des Mines de Paris, BP 87, F-91 003 Evry Cedex, France.

S. BROOKS

Central Electricity Research Laboratories, Kelvin Avenue, Leatherhead, Surrey KT22 7SE, UK.

M. CAILLET

Laboratoire d'Adsorption et Réaction de Gaz sur Solides, Equipe associée au CNRS No. 368, Ecole Nationale Supérieure d'Electrochimie et d'Electrométallurgie, BP 44, Institut National Polytechnique de Grenoble, 38401 Saint Martin d'Hères, France.

C. CARUGHI

FIAT (CRF), Orbassano, Italy.

J. F. CHRÉTIEN

Département de Génie Mécanique, Université de Technologie de Compiègne, BP 233, F-60 206 Compiègne, France.

G. L. COCCIA

Laboratori di Elettrochimica del Politecnico di Milano, Piazza Leonardo da Vinci, 32, I-20 133, Milano, Italy.

C. CODDET

Département de Génie Mécanique, Université de Technologie de Compiègne, BP 233, F-60 206 Compiègne, France.

J. C. COLSON

Laboratoire de Recherches sur la Réactivité des Solides associé au CNRS, Faculté des Sciences-Mirande, 21000 Dijon, France.

H.-J. DORN

Motoren- und Turbinen-Union, Postfach 50 06 40, D-8000 München 50, West Germany.

U. DUCATI

Laboratori di Elettrochimica del Politecnico di Milano, Piazza Leonardo da Vinci 32, I-20 133, Milano, Italy.

P. ELLIOTT

Corrosion and Protection Centre, UMIST, Manchester M60 1QD, UK.

E. ERDOES
Gebr. Sulzer AG, Metallkunde, Abt. 15, CH-8401, Winterthur, Switzerland.

P. C. FELIX
Brown Boveri & Cie, Abt. TCT, CH-5400, Baden, Switzerland.

J. M. FERGUSON
Central Electricity Research Laboratories, Kelvin Avenue, Leatherhead, Surrey KT22 7SE, UK.

S. FERRE
SNECMA, Division Métallurgie-Résistance, BP 81, F-91003 Evry Cedex, France.

E. FITZER
Institut für Chemische Technik der Universität Karlsruhe, Postfach 6380, D-7500 Karlsruhe, West Germany.

B. FÜRST
Brown Boveri & Cie AG, Zentrales Forschungslabor, Postfach 101332, D-6900 Heidelberg, West Germany.

A. GALERIE
Laboratoire d'Adsorption et Réaction de Gaz sur Solides, Equipe associée au CNRS No. 368, Ecole Nationale Superiéure d'Electrochimie et d'Electrométallurgie, BP 44, Institut National Polytechnique de Grenoble, 38401 Saint Martin d'Hères, France.

G. H. GESSINGER
Brown Boveri & Cie, Abt. TCT, CH-5400, Baden, Switzerland.

H. W. GRÜNLING
Brown Boveri & Cie AG, Postfach 351, D-6800 Mannheim 1, West Germany.

U. W. HILDEBRANDT
Brown Boveri & Cie AG, Zentrales Forschungslabor, Postfach 101332, D-6900 Heidelberg, West Germany.

D. R. HOLMES

Central Electricity Research Laboratories, Kelvin Avenue, Leatherhead, Surrey KT22 7SE, UK.

P. HUBER

Gebr. Sulzer AG, Metallkunde, Abt. 15, CH-8401, Winterthur, Switzerland.

P. J. HUNT

Department of Materials Science, Newcastle upon Tyne Polytechnic, Faculty of Science and Technology, Ellison Building, Ellison Place, Newcastle upon Tyne NE1 8ST, UK.

R. KRAFT

Kernforschungszentrum Karlsruhe, Institut für Material- und Festkörperforschung II, Postfach 3640, D-7500 Karlsruhe 1, West Germany.

M. LAMBERTIN

Laboratoire de Recherches sur la Réactivité des Solides associé au CNRS, Faculté des Sciences-Mirande, 21000 Dijon, France.

J. P. LARPIN

Laboratoire de Recherches sur la Réactivité des Solides associé au CNRS, Faculté des Sciences-Mirande, 21000 Dijon, France.

S. LEISTIKOW

Kernforschungszentrum Karlsruhe, Institut für Material- und Festkörperforschung II, Postfach 3640, D-7500 Karlsruhe 1, West Germany.

P. C. MARTINENGO

FIAT (CRF), Orbassano, Italy.

H.-J. MÄURER

Institut für Chemische Technik der Universität Karlsruhe, Postfach 6380, D-7500 Karlsruhe, West Germany.

D. B. MEADOWCROFT

Central Electricity Research Laboratories, Kelvin Avenue, Leatherhead, Surrey KT22 7SE, UK.

H. Meisel
Lehrstuhl für Metallurgie und Metallkunde, Technische Universität, Arcisstrasse 21, D-8000 München 2, West Germany.

R. Morbioli
SNECMA, Division Métallurgie-Résistance, BP 81, F-91003 Evry Cedex, France.

A. Moreau
Electricité de France, Traitement Industriel des Résidus Urbains, Usine d'Ivry, 43, rue Bruneseau, F-75 013 Paris, France.

A. R. Nicoll
Brown Boveri & Cie AG, Zentrales Forschungslabor, Postfach 101332, D-6900 Heidelberg, West Germany.

W. Nowak
Institut für Chemische Technik der Universität Karlsruhe, Postfach 6380, D-7500 Karlsruhe, West Germany.

R. Pichoir
Office National d'Etudes et de Recherches Aerospatiales (ONERA), F-92320, Chatillon, France.

E. Pott
Kernforschungszentrum Karlsruhe, Institüt für Material- und Festkörperforschung II, Postfach 3640, D-7500 Karlsruhe 1, West Germany.

A. Rahmel
DECHEMA, Postfach 97 01 46, D-6000 Frankfurt (M) 97, West Germany.

C. St. John
l'Ecole Nationale Supérieure des Mines de Paris, BP 87, F-91 003 Evry Cedex, France.

Kh. G. Schmitt-Thomas
Lehrstuhl für Metallurgie und Metallkunde, Technische Universität, Arcisstrasse 21, D-8000 München 2, West Germany.

K. SCHNEIDER

  Brown Boveri & Cie AG, Postfach 351, D-6800 Mannheim 1, West Germany.

C. G. STEVENS

  Central Electricity Research Laboratories, Kelvin Avenue, Leatherhead, Surrey KT22 7SE, UK.

K. N. STRAFFORD

  Department of Materials Science, Newcastle upon Tyne Polytechnic, Faculty of Science and Technology, Ellison Building, Ellison Place, Newcastle upon Tyne NE1 8ST, UK.

J. STRINGER

  Department of Metallurgy and Materials Science, University of Liverpool, P.O. Box 147, Liverpool L69 3BX, UK.

T. J. TAYLOR

  Corrosion and Protection Centre, UMIST, Manchester M60 1QD, UK.

G. WAHL

  Brown Boveri & Cie AG, Zentrales Forschungslabor, Postfach 101332, D-6900 Heidelberg, West Germany.

D. P. WHITTLE

  Department of Metallurgy and Materials Science, University of Liverpool, P.O. Box 147, Liverpool L69 3BX, UK.

1

Influence of the Composition of Fe–Cr–Al Ferritic Steels on the Corrosion Processes and Rates in Sulfur at High Temperature

J. P. LARPIN, M. LAMBERTIN and J. C. COLSON

Université de Dijon, France

ABSTRACT

High temperature corrosion (700–1000 °C) kinetics and processes in sulfur vapor or hydrogen sulfide were investigated for ferritic steels (Fe–Cr–Al). Observation and analysis of the reaction curves obtained allowed the various corrosion steps to be clarified. Formation of compact, protective layers either at the alloy/sulfide or the sulfide/gas interface occurred in the early stages of reaction, but subsequently most of the layers formed were microcrystalline and porous. The influence of the system parameters, temperature, pressure of sulfidizing agent and alloy composition, are reported in this chapter.

INTRODUCTION

The effects of chromium or aluminum additions on the corrosion of alloys in complex, industrial atmospheres, both oxidizing and sulfidizing, have been investigated by numerous authors (1–6). Some, however, devoted their efforts to more basic aspects, e.g. Strafford's work (7, 8) on Fe–Al alloys in sulfur vapor or H_2S/H_2 mixtures, and Mrowec's (9) and Nishida's (10, 11) work on Fe–Cr alloys in sulfur vapor at atmospheric pressure and high temperature. Sulfidation of ternary alloys was investigated by Simkovich (12) and of industrial chromium- and aluminium-containing steels in a sulfur vapor atmosphere by Mrowec (13, 14).

We have already reported the beneficial effects of aluminum additions to Cr-containing alloys for resistance to sulfidizing gases and have established that the corrosion rate decreases as the Al/Cr ratio increases (15).

1

EXPERIMENTAL

The kinetic studies were carried out in sulfur vapor and in H_2S/H_2 mixtures. A silica-spring MacBain type, continuous-recording thermobalance was used. For the sulfur vapor experiments, the whole device was placed in an enclosure kept at high temperature (200 °C) to obviate any condensation. The assembly has been described elsewhere (16).

TABLE 1

Composition of alloys

Alloy	C	Mn	Si	S	P	Ni	Cr	Al	N	Al%/Cr%
I	0.007	0.58	0.05	0.015	0.007	0.17	9.35	2.41	0.020	0.25
II	0.007	0.54	0.05	0.019	0.005	0.20	5.08	2.27	0.015 5	0.44
III	0.008	0.57	0.05	0.017	0.006	0.20	5.24	5.48	0.019	1.04
IV	0.006 5	0.54	0.07	0.015	0.005	0.20	3.10	6.37	0.025	2.05

The nature and composition of the scales formed were determined by X-ray diffraction. Sulfide scale morphology was observed by optical and scanning electron microscopy (Cambridge Stereoscan). An energy dispersive X-ray analyzer coupled to the latter allowed the various elements in the layers formed to be located.

The results reported in the present chapter are for four low-carbon steels containing small quantities of chromium and aluminum. Their composition is listed in Table 1.

The ferritic structure is obtained through hot rolling followed by annealing for 30 min at 850 °C and quenching in water. Coupons (10 × 10 × 2.5 mm), drilled to allow suspension in the thermobalance, were used. The samples were polished with 800 grit paper, washed with water and alcohol and then dried.

RESULTS

Kinetic Studies

(1) *Corrosion in sulfur vapor*

Specimens of alloys I, II, III and IV were exposed in a sulfur vapor pressure of 6.2×10^{-2} torr and at temperatures ranging from 700 °C to 1000 °C. The reaction curves, $\Delta m/S$ vs. $f(t)$ (Δm = specimen weight gain; S = specimen area) vary according to alloy composition in the early stages of reaction and then in all cases become linear (Figs. 1–3). Experimental activation energies deduced from the slopes of the linear parts of the curves varied from 13 to 22 kcal/Mol (Fig. 4) depending on the alloy.

(2) *Corrosion in H_2S/H_2 mixtures*
Specimens of alloy IV only were investigated in these conditions at constant temperature (794 °C) at gas mixture pressures from 40 to 210 torr. Mixture composition ($pH_2/pH_2S = 2$) was such that the sulfur partial pressure resulting from hydrogen sulfide dissociation was always 6×10^{-2} torr.

The reaction curves, $\Delta m/S$ vs. $f(t)$ in this case show a rapid initial rate followed by a steadily decreasing rate (Fig. 5). The transformed curves $(\Delta m/S)^2 = f(t)$ are linear (Fig. 6) throughout the latter stage, thus following Tamman's law.

Morphology and Composition of Scales
Sectioned and polished specimens were examined after treatment in sulfur vapor or H_2S/H_2 mixtures. For a number of experiments inert platinum markers were placed on the initial specimen surface to allow the examination of mass transfer processes.

The specimens used were exposed at 750, 850 and 950 °C in sulfur vapor ($p_{S_2} = 6.2 \times 10^{-2}$ and 1 torr) for periods from 1 h to 140 h or at 795 °C in various pressures of H_2S/H_2 mixtures (40–210 torr) for a few hours.

Three morphologies were exhibited depending on the reaction conditions, i.e. pressure, temperature, sulfiding agent and duration. In the early stages of reaction (when the $\Delta m/S$ vs. $f(t)$ curves are not yet linear and in the sulfur vapor reactions) a very thin, compact, inner layer covered by a thicker, microcrystalline layer was seen to form (Fig. 7). X-ray energy dispersive analysis showed high chromium and aluminum concentrations in the compact inner layer. The three main alloy components (Fe, Cr, Al) are present almost uniformly throughout the outer layer. The inner sublayer disappears more or less rapidly, depending on the alloy composition and temperature.

Subsequently, two different morphologies can be observed. When the $\Delta m/S$ vs. $f(t)$ curves are linear, only the microcrystalline layer growth is seen (Fig. 8a). Analysis of the composition of this layer showed the presence of iron, chromium and aluminum almost uniformly distributed (Fig. 8b–d). X-ray diffraction analysis revealed the existence of iron sulfide, $Fe_{1-x}S$ (pyrrhotite) and chromium sulfide (Cr_2S_3).

The presence of aluminum sulfide (Al_2S_3) could not be proved unambiguously, perhaps because of the low aluminum contents of the

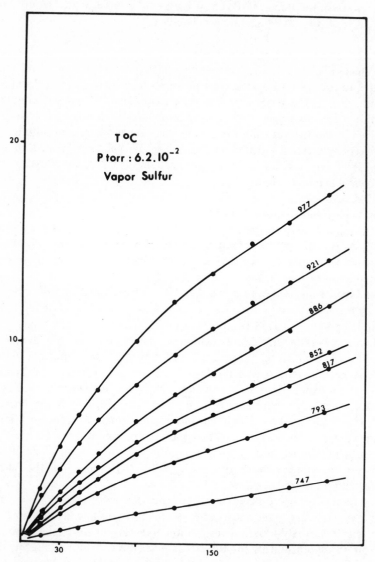

FIG. 1.   Alloy I: sulfidation kinetics in S_2 vapor. Ordinate: $\Delta m/S \, \mathrm{mg \, cm^{-2}}$.
Abscissa: time (min)

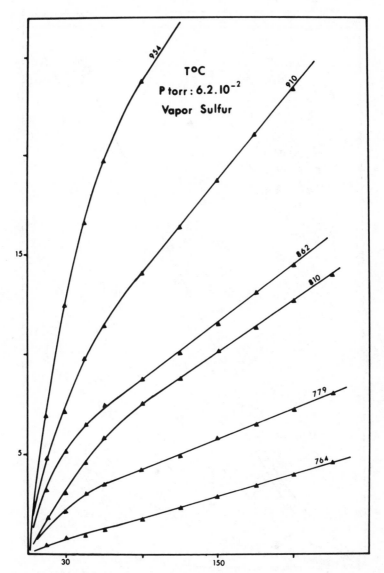

FIG. 2.   Alloy II: sulfidation kinetics in S_2 vapor. Ordinate: $\Delta m/S\,\mathrm{mg\,cm^{-2}}$.
Abscissa: time (min)

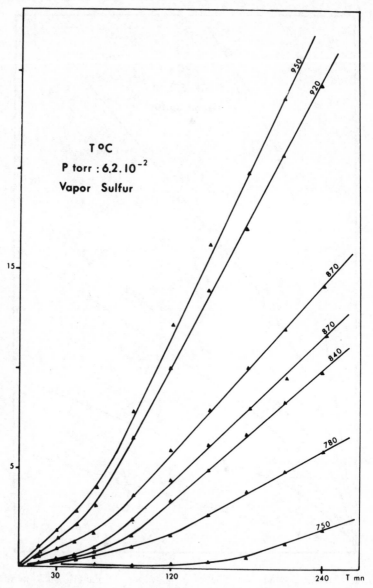

FIG. 3.   Alloy III: sulfidation kinetics in S_2 vapor. Ordinate: $\Delta m/S \, \text{mg cm}^{-2}$.
Abscissa: time (min).

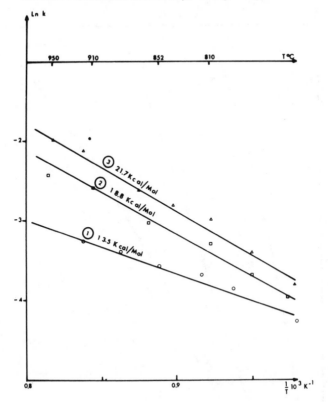

FIG. 4.   Alloys I, II and III: activation energy

initial alloys and the instability of this phase in air (17). In some cases, the diffraction patterns of a complex sulfide, $FeCr_2S_4$ (daubreelite), were recorded.

The inert markers always remained at the outer sulfide/gas interface whatever the thickness of the layer formed (Fig. 9).

Following much longer treatments (always over 15 h in sulfur vapor) two distinct, thick layers were noticed (Fig. 10): an inner, finely crystallized layer similar to that described previously and an outer, basalt-like one composed of iron sulfide (pyrrhotite, $Fe_{1-x}S$) only. The latter compact layer occurs much more rapidly with exposure in H_2S/H_2 mixtures than in sulfur vapor. Its growth coincides with the parabolic part of the $\Delta m/S$ vs. $f(t)$ curves (Fig. 5). The fact that the 'basaltic' layer growth also corresponds to a $\Delta m/S$ vs. $f(t)$ parabolic curve when corrosion occurs in

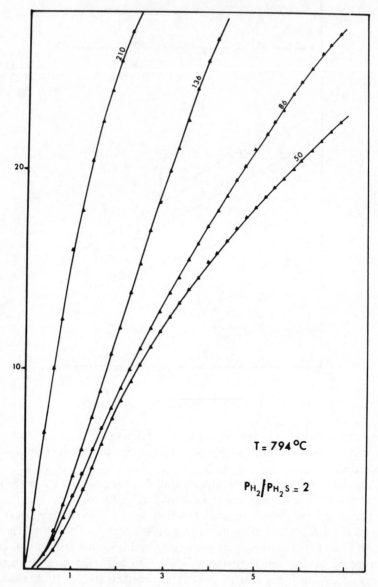

FIG. 5. Alloy IV: sulfidation kinetics in H_2S/H_2 mixtures. Ordinate: $\Delta m/S \, \mathrm{mg \, cm^{-2}}$. Abscissa: time (h). Curves labelled with total gas pressure (torr).

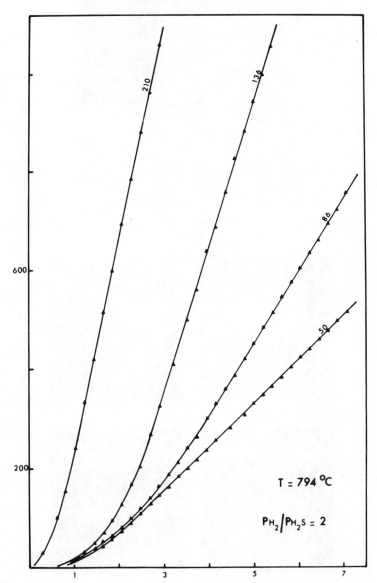

FIG. 6. Alloy IV: transformed curves. Ordinate: $[\Delta m/S]^2 \, \mathrm{mg}^2 \, \mathrm{cm}^{-4}$. Abscissa: time (h). Curves labelled with total gas pressure (torr).

FIG. 7. Cross-section of scale on alloy III sulfidized for 60 min at 850 °C in sulfur vapor

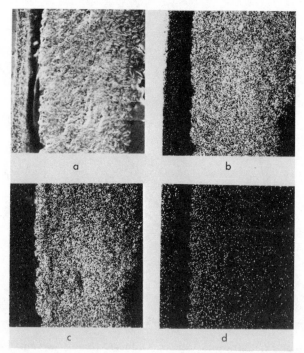

FIG. 8. Cross-section of scale on alloy II sulfidized for 5 h at 950 °C in sulfur vapor; (a) electron image; (b) Fe X-ray image; (c) S X-ray image; (d) Cr X-ray image

FIG. 9. Marker experiment on scale on alloy II sulfidized for 50 h at 850 °C in sulfur vapor

sulfur vapor could also be shown (Fig. 11). After sulfidation, the inert platinum marker remains at the interface between the two corrosion layers (Fig. 10).

The results obtained for the various specimens as a function of the experimental conditions (sulfiding agent, temperature, pressure, duration) are summarized in Table 2.

It is to be noted that formation of the outer compact layer is favored by an increase of the sulfidizing agent pressure (sulfur or hydrogen sulfide) and

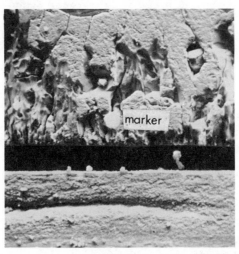

FIG. 10. Marker experiment on scale formed on alloy II sulfidized for 100 h at 850 °C in sulfur vapor

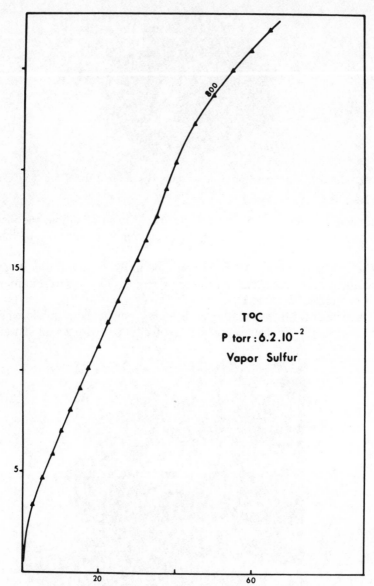

FIG. 11.   Alloy II: sulfidation kinetics in S_2 vapor showing the three stages of the reaction. Ordinate: $\Delta m/S \, \mathrm{mg \, cm^{-2}}$. Abscissa: time (h)

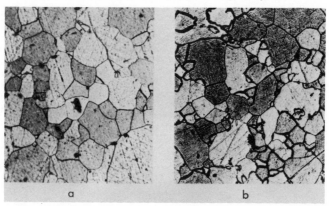

Fig. 12.   Top view of surface of alloy II: (a) initial alloy; (b) after sulfidation for 5 h
at 750 °C in sulfur vapor

by a decrease in temperature. In similar conditions, it was noticed that the
time needed by the outer compact layer to occur depends on the
composition of the specimen.

As the inert marker position showed that the thickening of the
microcrystalline layer always occurs through inward movement of
sulfidant, we thought it worth while examining the metal/sulfide interface
more closely. After corrosion scale removal and slight polishing of the
specimen the grain boundaries and subgrain boundaries were seen to be
decorated with corrosion products (Fig. 12). Sulfur penetration into the

TABLE 2
Sulfidation conditions

Temperature ($°C$)	Sulfiding agent pressure (torr)	Alloy	Number of layers	Duration (h)
$700 < T < 1\,000$	$S_2 - 6.2 \times 10^{-2}$	I–IV	1	<5
750	$S_2 - 6.2 \times 10^{-2}$	II	2	16
850	$S_2 - 6.2 \times 10^{-2}$	II	1	25
850	$S_2 - 6.2 \times 10^{-2}$	II	2	50–140
850	$S_2 - 1$	II	1	10
850	$S_2 - 1$	II	2	20
950	$S_2 - 6.2 \times 10^{-2}$	II	1	50
850	$S_2 - 6.2 \times 10^{-2}$	IV	1	50
795	H_2S/H_2 40	IV	1	7
795	H_2S/H_2 50 to 210	IV	2	3–7

grain boundaries is about $60\,\mu m$ and aluminum and chromium also accumulate there.

DISCUSSION

From the results obtained, three successive stages may be assumed in the corrosion scale growth; they are shown in Fig. 13.

Stage A. This corresponds to the initial formation of the thin compact

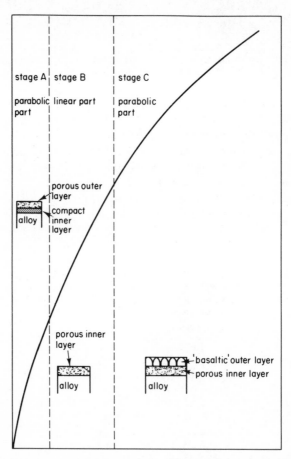

FIG. 13.   Correlation of the three stages in the reaction curve with the scale microstructures. Ordinate: $\Delta m/S$ mg cm^{-2}. Abscissa: time (h)

layer and its preservation for some time, even while the microcrystalline layer is growing (Fig. 7). Formation of the first compact sublayer must correspond to penetration of sulfidant to the alloy and the preferential formation of sulfides from those metals having the highest affinity for sulfur. Some iron sulfide also appears; this is due to the high concentration of iron in the alloy. Analysis of the weight gain curves over this stage proves difficult, as the shapes obtained depend on the temperature and pressure conditions and the alloy composition. This non-reproducibility is possibly to be ascribed to the fact that the overall corrosion rate is determined by several processes; in particular the formation of the outer porous layer may result from the break-up of this first layer by a number of different processes. Our observations, however, tend to substantiate an inner layer corrosion process involving the following steps: dissociation of initially formed FeS, sulfur dissolution at the alloy grain boundaries followed by chromium and aluminium sulfide nucleation (Fig. 14).

Iron released through FeS decomposition is sulfidized once again at the interface between the two layers. The large volume ratios corresponding to the formation of various sulfides ($\Delta_{Al \to Al_2S_3} = 3.6$, $\Delta_{Cr \to Cr_2S_3} = 3.9$, $\Delta_{Fe \to FeS} = 2.7$) generate mechanical stresses causing the outer layer to break up, thus inducing porosity which permits gaseous diffusion of sulfur vapor or hydrogen sulfide. This process is repeated, ensuring the continued growth of the porous outer layer whilst allowing the thin compact layer at the sulfide/alloy interface to remain.

Stage B. This corresponds to the linear part of the $\Delta m/S$ vs. $f(t)$ curves. It shows the previously described porous layer growth, which consists of the three sulfides $Fe_{1-x}S$, Cr_2S_3 and Al_2S_3, homogenously distributed and probably present as the thiocompounds, $FeCr_2S_4$, $FeAl_2S_4$ and $FeCr_xAl_{2-x}S_4$. Sulfur or hydrogen sulfide can always diffuse through the pores towards the alloy/scale interface. From the kinetic standpoint, the progress of the reaction shows that gas diffusion through the porous layer is not the rate-determining process but that the critical process is some localized phenomenon such as the sulfur dissolution–sulfide nucleation step in the alloy close to the sulfide/metal interface.

Stage C. This corresponds to growth of an outer compact layer of iron sulfide, $Fe_{1-x}S$, while the inner porous layer continues to thicken. The observation that the marker remains at the porous layer/compact layer interface (Fig. 10) indicates that the latter grows by diffusion of iron ions towards the outer sulfide/gas interface, which confirms the parabolic $\Delta m/S$ vs. $f(t)$ curves and the transformation from linear kinetics (Figs. 5, 6 and 11). The formation of this compact layer over the porous one may be

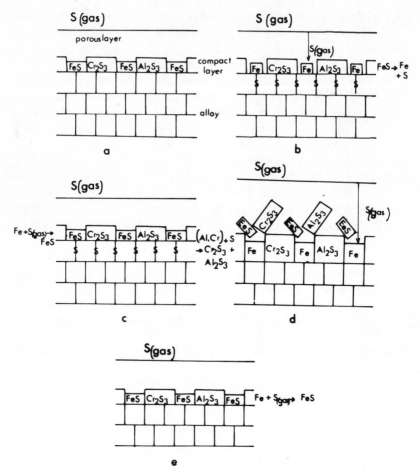

FIG. 14. Sulfidation mechanism during the first stage of the reaction

explained by recrystallization of the latter, especially as it contains at least 85% iron sulfide. From the experimental results it is difficult to determine the causes of the more or less rapid formation of the iron sulfide outer layer. At low sulfur vapor pressures the outer layer grows only after rather long periods (more than 15 h), whereas it occurs very rapidly in hydrogen sulfide pressures above 30 torr (Table 2).

Two factors are of greater basic importance than alloy composition— corrosion temperature and reactant gas pressure.

Actually, layer recrystallization is explicable in terms of the occurrence of

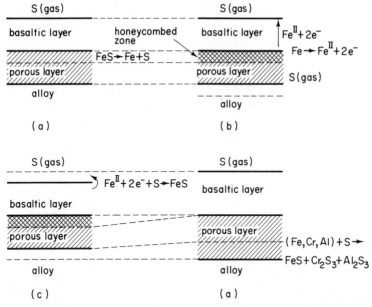

FIG. 15. Sulfidation mechanism during the third stage of the reaction

stresses due to the volume changes associated with sulfide formation. It is, therefore, favored by a decrease in corrosion rate and temperature. As soon as a continuous scale can form, it grows, as on pure metals, by iron diffusion through the outer compact layer.

Chromium and aluminum sulfidation (as well as of part of the iron) proceeds by internal growth (Fig. 15). Hence, a deep honeycombed area should form near the compact layer/porous layer interface, as with the sulfidation of pure metals (18). This does not occur. This may be due to the plasticity of the microcrystalline layer and the large volume changes associated with the formation of the various sulfide phases.

CONCLUSIONS

The various stages in the corrosion of Fe–Cr–Al ferritic alloys at high temperature were determined by analyzing the reaction curves and observing the layers formed. As shown earlier, the processes involved are complex and result in the formation of corrosion scales with various morphologies.

Formation of compact, protective layers, either at the alloy/sulfide interface or at the sulfide/gas interface, occurs in the early stages of corrosion and even subsequently, but most of the layers formed are microcrystalline and porous.

Every system parameter, in particular the pressure of the sulfidizing agent, affects the nature of the layers formed. At this stage of our research, however, it does not yet seem possible to induce the formation and growth of a compact, protective film which would improve the behavior of the material in a sulfidizing atmosphere.

REFERENCES

1. S. K. Verma, D. P. Whittle and J. Stringer (1972). *Oxid. Met.*, **5**(3), 169.
2. J. D. McCoy (1974). *Mat. Perf.*, **13**(5), 19.
3. H. Doi, T. Onoda and Y. Harada (1972). *Proc. Gas Turbine and Fluids Eng. Conf. Prod. Show* (San Francisco, March 1972), ASME.
4. H. Morrow III, D. L. Sponseller and E. Kalns (1974). *Met. Trans.*, **5**, 673.
5. P. W. Rice, R. A. Beverly, C. E. Ney and B. R. Kee Ney (1973). *Mat. Perf.*, **12**(10), 25.
6. J. D. McCoy and F. B. Hamel (1971). *Mat. Perf.*, **10**(4), 17.
7. K. N. Strafford and R. Manifold (1969). *Oxid. Met.*, **1**(1), 21.
8. K. N. Strafford and R. Manifold (1969). *Corros. Sci.*, **9**, 489.
9. S. Mrowec, T. Walec and T. Werber (1969). *Oxid. Met.*, **1**(1), 93.
10. T. Narita and K. Nishida (1973). *Oxid. Met.*, **6**(3), 157.
11. K. Nishida, K. Nakayama and T. Narita (1973). *Corros. Sci.*, **13**, 759.
12. P. O. Zelanko and G. Simkovich (1973). *Oxid. Met.*, **8**(5), 343.
13. S. Mrowec, T. Werber and J. Podhorodecki (1968). *Corros. Sci.*, **8**, 815.
14. S. Mrowec, S. Tochowicz, T. Werber and J. Podhorodecki (1968). *Corrosion*, **7**, 697.
15. Y. Morel, J. P. Larpin, M. Lambertin and J. C. Colson (1976). Table Ronde Franco-Polonaise, Dijon (September).
16. M. Lambertin and J. C. Colson (1972). *Bull. Soc. Chim.*, **2**, 561.
17. Y. Trambouze (1961). In *Nouveau Traité de Chimie Minérale*, Vol. IV, Ed. P. Pascal, Masson, Paris, p. 610.
18. M. Lambertin and J. C. Colson (1973). *Oxid. Met.*, **7**(3), 163.

DISCUSSION

I. Lefever (N.V. Bekaert S.A., B-8550 Zwevegem, Belgium): Did you investigate the influence of the surface condition (roughness) on the corrosion rate?

Reply: The tests have been carried out on polished samples. There is certainly an influence of the surface roughness but in the present investigation we did not consider that aspect.

I. Lefever: Can you give us an idea of the thicknesses of the porous and of the compact layer and also the corrosion penetration depth along the grain boundaries?

Reply: The thickness of the compact layer: a few microns. The thickness of the porous layer: up to a few tens of microns. The penetration depth along the boundaries: 50–60 microns.

M. Dixmier (Commissariat à l'Energie Atomique, France): I should like to make a remark and to ask a question. First, I think it is always unsafe to decide on the choice of materials after tests of only a few hours; you have made it very clear that after a sufficiently long exposure, at high temperatures, the behavior of Fe–Cr–Al alloys is poor. It must be emphasized that your results, which agree perfectly with ours, totally disagree with other technical literature which recommends such alloys for use at 500 °C when sulfur vapor is present.

My question is, therefore, do you think that tests in these conditions, to obtain an alloy with suitable iron, chromium and aluminium contents, will in fact lead to a safe and resistant alloy in high temperature sulfur?

Reply: Of course, we would like to answer *yes* to your question. Unfortunately our research was not taken far enough to be able to do this. However, even if these particular alloys do not fulfil the characteristics required of them, some of their properties will perhaps enable us to find a satisfactory solution to this problem.

R. Hales (BNL, Berkeley, USA): We have found that small amounts of sulfur in CO_2 seriously affect the oxidation of 18 Cr austenitic alloys; for example, types 316 and 321 are adversely affected. Does the addition of aluminum improve the oxidation resistance of austenitic steels as well as of ferritic steels?

Reply: From our results we can conclude that the aluminum addition improves the sulfidation behavior of ferritic Fe–Cr–Al alloys, at least for a longer or shorter period at the beginning of the exposure. We have no results concerning the *oxidation* behavior of austenitic or ferritic steels. But the results obtained by numerous authors prove that sufficient amounts of aluminum and chromium to form a continuous Al_2O_3 layer improve the oxidation resistance of steels.

A. Rahmel (DECHEMA, Frankfurt, Germany): Internal sulfidation should not occur if the Al content of the alloy is high enough to form a dense and complete Al_2S_3 film on the surface. Did you observe this with the alloys tested?

Reply: I agree with your assumption: the internal sulfidation should not take place if a dense and complete Al_2S_3 film is formed on the surface, but we never observed this. The conditions to obtain such a layer are certainly similar to those which are required in the case of oxidation. If we use the results obtained in the latter case, the concentrations of aluminum and particularly of chromium in our alloys are not sufficiently high to form a protective Al_2S_3 layer and so avoid internal sulfidation.

2

Comparative Studies of the High Temperature Oxidation of Four Austenitic CrNi Steels in Helium with Water Vapor and Hydrogen Additions

S. Leistikow, R. Kraft and E. Pott

Institut für Material- und Festkörperforschung, Karlsruhe, West Germany

ABSTRACT

The long-term corrosion of potential fuel element cladding materials for a helium-cooled fast breeder reactor was tested in helium at high temperatures under steam generator leak conditions. For this purpose, four different austenitic stainless steels (material Nos. 1.4876, 1.4970, 1.4981 and 1.4988), known for their good properties in the sodium-cooled fast reactor environment, were selected and exposed at 800°C to a helium atmosphere containing 80–360 µatm hydrogen and 3.75–8.1 µatm water vapor. Oxidation kinetics were studied by gravimetry and oxidation product morphology by metallography and electron microanalysis.

In accordance with earlier results, it was shown that Nb-bearing steels have excellent properties under these conditions, forming only thin surface oxide layers and no internal oxidation products. Ti-, Al-bearing steels, in contrast formed thicker oxide layers, underwent internal attack and showed changes in surface composition by formation of diffusion boundary layers. Experimental results and discussions are given on oxidation rate laws, dependency of oxidation rates on partial pressure of oxidizing species and oxidation potential, and high temperature mechanical property changes during long-term exposure.

INTRODUCTION

The recommendations of an OECD expert commission convened in 1973 stated that potential fuel element cladding materials to be used in a helium-cooled fast breeder reactor should undergo long-term corrosion studies in helium, containing known concentrations of water vapor and hydrogen, in order to simulate leakages occurring in the steam generator.

These test conditions were to be applied to those CrNi steels proposed by German experts as a result of extensive experience gained in the DEBENELUX project of a sodium-cooled fast breeder reactor, although—with the exception of AISI 316 stainless steel—they had not yet been subjected to testing in contaminated helium.

In these studies special attention had to be paid to the following corrosion effects:

(1)   Kinetics of surface layer formation in general, chemical composition, and adhesive strength of the surface layer.

(2)   Kinetics of the intergranular oxidation (internal attack) and formation of a diffusion boundary layer in the metallic surface, chemical composition of the oxidation products and boundary layers, and possible effects on the mechanical behavior of the materials.

The latter seemed to be particularly important for the titanium- and aluminum-bearing high temperature materials, such as AISI 321 (1–7), 12R72HV (8), Incoloy 800 (9) and PE 16 (10), since they had been found to be sensitive to intergranular oxidation in a series of previous studies.

The programme outlined above called for the construction of a special apparatus to allow long-term high temperature reactions to be performed in defined gas mixtures at atmospheric pressure. Thus, in addition to the experimental results, the successful operation of a very sensitive metering technique will be reported.

EXPERIMENTAL

Conduct of Experiment
Materials tested
Four different austenitic CrNi steels (Table 1) were included in the investigations. The steels differed from each other as follows:

The nickel content varied from 13.6 to 31.6 % and the chromium content from 14.9 to 20.5 %. However, interest should concentrate on the contents of the elements aluminum, titanium and niobium, which are less than 1 %.

Within the framework of R & D investigations related to a sodium-cooled fast breeder reactor, the CrNi stainless steels Nos. 1.4971, 1.4988

TABLE 1
Chemical composition of CrNi stainless steels (wt%)

Material No.	C (%)	Si	Mn	Cr	Ni	Nb	Ti	Mo	Al	P	S	Co	Cu	B	N_2	Fe
1.4981	0.06	0.61	1.3	17.6	16.7	0.80	—	1.8	—	0.020	0.024	—	—	—	—	Bal.
1.4988	0.08	0.29	1.1	16.0	13.6	0.87	—	0.8	—	0.017	—	1.34	—	—	—	Bal.
1.4970	0.10	0.60	1.7	14.9	15.2	—	0.48	1.2	—	0.009	0.008	—	—	0.009	0.006	Bal.
1.4876	0.05	0.50	0.7	20.5	31.6	—	0.40	0.2	0.37	—	0.003	0.13	0.10	—	—	Bal.

and 1.4970 were subjected to extensive testing. Steel No. 1.4970 emerged as the material having the best properties for use in breeder reactors. CrNi stainless steel No. 1.4876 (Incoloy 800), a familiar superheater material, was used for comparison.

Prior to the corrosion tests, flat tensile specimens having a length of 53.3 mm, a width of 4.2 mm, and a thickness of 0.5–1.2 mm were punched out from annealed steel sheets. They were electropolished, cleaned, dried and weighed before use. In this way it was possible to investigate both the oxidation kinetics and the influence of oxidation on material mechanical properties at the temperature of operation.

Although this technique of post-exposure mechanical investigation does not provide information on the actual material behavior under the combined effects of oxidation and mechanical creep deformation, it seems to be reasonable for evaluating the influence exerted by oxide penetration on the mechanical properties.

Gas mixtures used in the tests
The impurities metered into high purity helium were varied in the tests, as shown in Table 2. To convert to the partial pressure (μatm) of metered impurities prevailing in the system, their concentrations (vpm) have to be multiplied by the system pressure (atm).

TABLE 2

Content of impurities		Ratio
H_2 (vpm)	H_2O	pH_2/pH_2O
53	5.4	10
126	2.7	50
242	2.5	100

For a pH_2/pH_2O ratio equal to 100, thermodynamic considerations indicate (Fig. 1) that only the oxides of the elements aluminium, titanium, niobium, manganese and chromium are formed while the oxides of the elements molybdenum, iron, cobalt and nickel are unstable.

Test apparatus
When the facility for corrosion studies (Fig. 2) on metallic materials in contaminated helium was built, the main problem concerned the controlled dosage of the desired low impurity concentrations, particularly the water

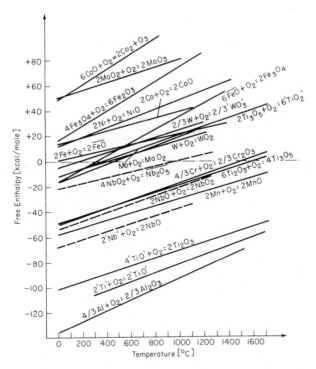

FIG. 1.   Free enthalpy ΔG_0 of formation of metal oxides at $pH_2/pH_2O = 100/1$ as function of temperature (after Huddle)

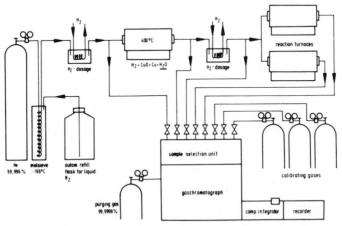

FIG. 2.   Experimental facility for high temperature oxidation reactions in helium containing dosed amounts of hydrogen and water vapour

content. Whereas helium can be bought with a multitude of gaseous impurities, it is not possible to keep, for extended periods of time, gases with a constant water content since the concentration undergoes temporal variations by alternating condensation and evaporation. Besides, conventional control valves, even of a most sophisticated design, are not capable of feeding the desired small amounts of hydrogen in the vpm range into the relatively small gas volumes required for the test sections.

The water-injection method used is based on the temperature-controlled diffusion of hydrogen through high purity nickel, followed by oxidation of the hydrogen with copper oxide at 400 °C to give water. In the test apparatus, high purity helium undergoes a secondary purification via a molecular sieve column cooled by liquid nitrogen. An automatic refilling device continuously supplements the liquid nitrogen level of the dewar in which the molecular sieve column has been placed. The helium, having undergone secondary purification, subsequently flows through a small stainless steel tank into which an approximately 10 cm-long helical nickel capillary (2 mm × 0.2 mm) is fixed, soldered directly on one side and by a ceramic insulator on the other side. The nickel helix, exposed to a pressure of 2 atm hydrogen in its interior, is directly heated by a variable ratio transformer connected to a voltage stabilizer (40 A; 1.5 V). In this way, the temperature of the nickel helix and, consequently, the hydrogen diffusion rate can be perfectly adjusted via the variable ratio transformer, so that setting the hydrogen concentration between 0 and about 300 vpm does not raise problems. A gas chromatograph with an automatic sampling device supplied by L'Air Liquide continuously records the hydrogen content and, in order to give an immediate indication of possible leakages, the nitrogen and oxygen contents in addition. It also allows the measurement of impurities such as CH_4, CO and CO_2.

The hydrogen-bearing helium is introduced into a copper oxide filled quartz tube heated up to 400 °C in which the hydrogen undergoes quantitative reaction to water. The hydrogen concentration is controlled by the gas chromatograph.

To set the desired water and hydrogen ratios the respective amounts of hydrogen are metered in, using a second hydrogen metering device also controlled by a gas chromatograph. The gas mixture thus prepared is subsequently brought via heated, very short stainless steel tubes into two parallel connected test sections heated by resistance furnaces, into which the specimen sheets are introduced on quartz slides. After it has left the heated test section, the gas composition is again monitored in a gas chromatograph.

Tests performed

In the investigations on the four CrNi steels, both the absolute concentrations of water vapor and hydrogen, and also the oxidation potential, were varied in the manner already mentioned. The tests were performed under isothermal conditions, each over 2500 h at 800 °C. Every 500 h the tests were interrupted and the specimens weighed. After termination of the tests final weighing was performed.

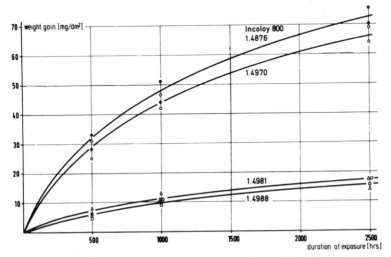

FIG. 3. Weight gain of various CrNi stainless steels as function of time of exposure (800 °C, 1.5 atm, $pH_2/pH_2O = 126/2.7$)

As an example, Fig. 3 shows the measured weight gains as a function of time, of the four CrNi steels when tested for 2500 h at 800 °C in helium containing 2.7 vpm H_2O and 126 vpm H_2.

Test Results

Results of gravimetry

The gravimetric experiments yielded an oxygen uptake which was relatively high in the case of the two Ti-, Al-containing steels. It was also independent of the environmental changes in water vapor (2.5–5 vpm), hydrogen content and their ratio (10, 50 or 100), indicative of the oxidation potential, and could be approximated in its weight gain–time relationship by parabolic rate laws. After 2500 h for both steels, Nos. 1.4970 (Fig. 4) and 1.4876 (Fig. 5), equal weight gains of about 70–80 mg/dm², corresponding to 4.6–5.3 μm oxide layer thickness, were measured.

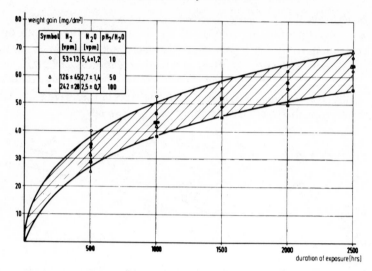

FIG. 4. Weight gain of CrNi stainless steel No. 1.4970 as function of time of exposure at various oxidation potentials (800 °C, 1.5 atm, $pH_2/pH_2O = 53$–$242/$ 2.5–5.4)

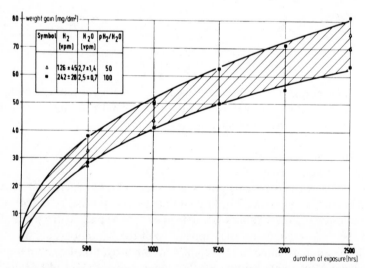

FIG. 5. Weight gain of CrNi stainless steel No. 1.4876 as function of time of exposure at various oxidation potentials (800 °C, 1.5 atm, $pH_2/pH_2O = 126$–$242/$ 2.5–2.7)

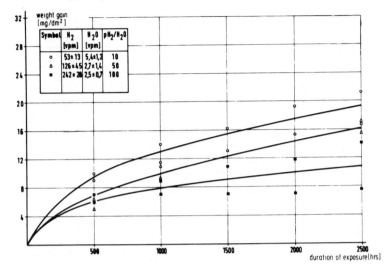

Fig. 6. Weight gain of CrNi stainless steel No. 1.4988 as function of time of exposure at various oxidation potentials (800 °C, 1.5 atm, pH₂/pH₂O = 53–242/ 2.5–5.4)

The two niobium-containing steels, Nos. 1.4988 (Fig. 6) and 1.4981 (Fig. 7), exhibited higher corrosion resistance. After 2500 h, the relatively low oxygen uptake even showed a slight reduction from 19 to 11 mg/dm², corresponding to 1.3 to 0.8 μm oxide layer thickness, when the above-mentioned environmental changes to lower water vapor and higher hydrogen contents were performed, thus changing the weight gain–time relationship from parabolic to cubic functions.

A bar chart (Fig. 8) summarizes the weight increases of the steels after 2500 h at 800 °C. For the steels Nos. 1.4981 and 1.4988 a clear tendency can be seen towards a smaller weight increase with lower water vapor content and lower oxidation potential, respectively. For the two titanium-bearing steels such a tendency cannot be clearly detected within the scatter band.

In comparison, high temperature corrosion tests performed on the Incoloy 800 alloy in superheated steam at atmospheric pressure yielded, after 2500 h, approximately double the weight increase obtained in the experiments performed specifically in contaminated helium.

Results of metallography
The metallographic investigations produced the following picture of the

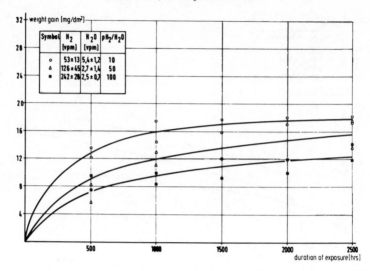

FIG. 7. Weight gain of CrNi stainless steel No. 1.4981 as function of time of exposure at various oxidation potentials (800 °C, 1.5 atm, $pH_2/pH_2O = 53–242/2.5–5.4$)

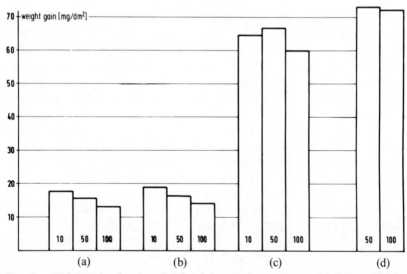

FIG. 8. Weight gain of various CrNi stainless steels at various oxidation potentials (2500 h, 800 °C, 1.5 atm, $pH_2/pH_2O = 10, 50, 100$). Material No.: (a) 1.4981, (b) 1.4988, (c) 1.4970, (d) 1.4876

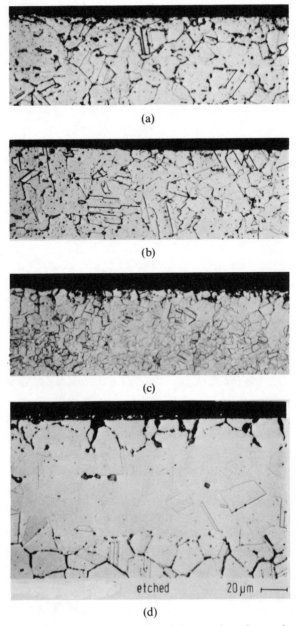

FIG. 9. Cross-section of various CrNi stainless steel surfaces after oxidation (2500 h, 800 °C, 1.5 atm, $pH_2/pH_2O = 126/2.7$). Material No.: (a) 1.4981, (b) 1.4988, (c) 1.4970, (d) 1.4876

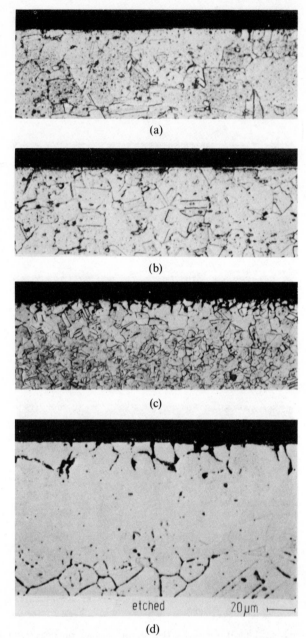

(a)

(b)

(c)

etched   20 μm ⊢——⊣

(d)

FIG. 10. Cross-section of various CrNi stainless steel surfaces after oxidation (2500 h, 800 °C, 1.5 atm, $pH_2/pH_2O = 242/2.5$). Material No.: see Fig. 9

different materials after corrosion in helium containing 2.5 vpm H_2O and 125 vpm H_2 for 2500 h at 800 °C (Figs. 9 and 10):

(1) For the niobium-stabilized steels Nos. 1.4981 and 1.4988 a continuous thin oxide layer was formed corresponding to the very low oxygen uptake. There was practically no oxide penetration.

(2) For material No. 1.4970 very pronounced oxide penetration (20 μm) occurred under different test conditions but with the same durations. The thickness of the chromium-depleted layer was about 35 μm and that of the oxide layer about 4 μm.

(3) For material No. 1.4876 the oxide layer roughly corresponded to that of material 1.4970. In contrast, oxide penetration (28 μm) and the chromium-depleted layer (80 μm) were much more pronounced.

Results of microanalysis

Using the scanning electron microscope, X-ray spectra of different portions of the structure were taken on the cross-sections. Microanalytical investigations of material No. 1.4876, which had the maximum oxidation rate showed that the principal elements present in the oxide layer were aluminum, titanium, chromium and manganese (Fig. 11).

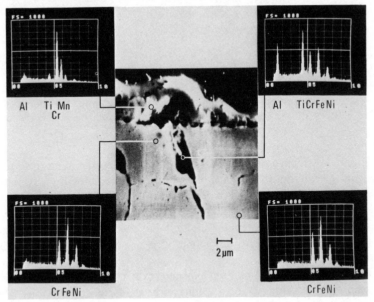

FIG. 11. Distribution of various elements in the oxidized surface of Cr Ni stainless steel No. 1.4876 after oxidation (2500 h, 800 °C, 1.5 atm, $pH_2/pH_2O = 126/2.7$)

34 S. Leistikow, R. Kraft and E. Pott

The intergranular corrosion products consisted primarily of aluminum and titanium but also contained chromium, iron and nickel. According to expectations, chromium, iron and nickel were detected in the metallic matrix.

For the corroded surface of steel No. 1.4970, a step-by-step analysis of the diffusion boundary layer from the metallic matrix to the oxide was performed (Fig. 12). A 3 μm × 3 μm area was scanned. In the diffusion

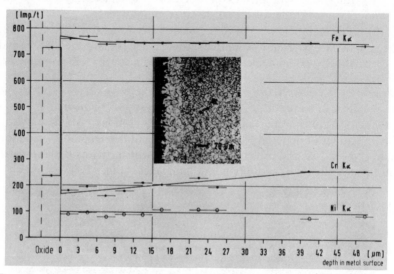

FIG. 12. Concentration gradients of the major elements in the diffusion zone of CrNi stainless steel No. 1.4970 after oxidation (2500 h, 800 °C, 1.5 atm, pH$_2$/pH$_2$O = 53/5.4)

boundary layer, chromium appears to be slightly depleted, and this can be seen on the micrograph of the etched material as a zone approximately 34 μm deep. For iron, the concentration increases slightly towards the oxide. In the oxide, the iron concentration decreases considerably while, at the same time, the chromium content increases. Nickel is uniformly distributed over a band in the metal and is present in the oxide in trace amounts only.

The extent of intergranular oxide penetration in material No. 1.4970 can be seen on the scanning electron micrographs of etched and non-etched specimens shown in Fig. 13.

Microhardness measurements in the diffusion boundary layers of the two

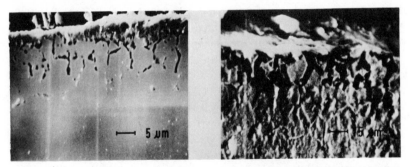

FIG. 13. Cross-section of oxidized CrNi stainless steel No. 1.4970 surface (2500 h, 800 °C, 1.5 atm, $pH_2/pH_2O = 53/5.4$); left: unetched, right: etched

titanium-bearing steels showed hardness values reduced by 10 % compared with the matrix, only in the case of material No. 1.4970.

Measurements of possible changes in mechanical properties were made only for material Nos. 1.4970 and 1.4876, as these were showing intergranular oxidation. For material No. 1.4970, a drop of the yield strength from 17 to 14 kPa/mm² and a decrease in true plastic strain from 19 to 15 % were found. For material 1.4876, the yield strength fell from 13.5 to 9 kp/mm² and true plastic strain from 17 to 15.5 %. The decrease in these two parameters relative to those for the unexposed material was greater for oxide-covered specimens than for descaled specimens.

SUMMARY

The most significant results of the experimental studies were the following:

(1) A solution to the problem of metering in impurities over extended test periods.

(2) The detection of a group of CrNi steels stable at 800 °C and of a group which would supposedly be susceptible to damage due to oxide penetration to a significant depth. The magnitude of this damage cannot easily be calculated in advance and may even increase under operating stress.

Obviously the intergranular oxidation and oxide penetration is not due to the different Cr and Ni contents of the steels investigated, but to the presence of alloying elements such as Al and Ti in concentrations less than 1 %. This is supported by the metallographic findings which show uniform oxide layers on steels free from these elements (Nos. 1.4981 and 1.4988), while the steels containing these elements (Nos. 1.4970 and 1.4876) showed regular oxidation of the surface and intergranular oxidation.

According to thermodynamic estimates, only the oxides of the alloying elements Cr, Mn, Ti, Al and Si should be formed at the selected oxidation potentials corresponding to $pH_2/pH_2O = 10 \rightarrow 100$, and these elements were in fact detected by microanalysis in the surface layer (Cr_2O_3) and in the internally oxidized parts of the structure (TiO_2, Al_2O_3).

In contrast, the kinetics of the oxidation reaction, derived from the increases in weight as a function of time, can be related to the concentration of water vapor and the oxidation potential, respectively, as parameters. The analysis of the rate law yielded a largely parabolic dependence on time which allows us to conclude that the uptake of oxygen is diffusion controlled. The oxygen is taken up, on the one hand, by outward diffusion of chromium ions forming an external oxide layer and, on the other hand, by inward diffusion of oxygen ions preferably along the grain boundaries while forming intergranular oxidation products. It is interesting that for Nb-bearing steels the oxygen uptake was shown to slow down markedly with decreasing water vapor content and increasing hydrogen content, i.e. with lower oxidation potential. This is more probably explained by the formation of a more protective oxide composition at the lower oxidation potentials rather than by a concentration-dependent transport of water vapor to the oxidizing metal surface as the oxidation rate determining step.

The parabolic time dependence of the weight increase also indicated that the penetration of the intergranular oxide and the growth of the chromium-depleted layer were diffusion controlled. The maximum values of penetration and depleted layer formation, respectively (No. 1.4876: $28/80\,\mu m$, No. 1.4970: $20/35\,\mu m$), reach a technically significant order of magnitude since, in practice an extrapolation must be made to predict the damage after a service life of at least three years; only a 350–$400\,\mu m$ wall thickness for the cladding tube can be tolerated. The chemical changes in the diffusion depleted layer correspond to those occurring during the oxidation of cold-worked materials in steam at low temperatures: depletion in chromium and manganese, enrichment in nickel and iron. The depletion in chromium is accompanied by the dissolution of chromium carbides segregated in the grain boundaries, as proved by the reduction in microhardness.

The observed changes in the high temperature mechanical properties (reduction of yield strength and of true plastic strain), particularly for the oxide-covered tensile specimens of the Ti- and Al-bearing materials, are not statistically significant, so that interpretations and predictions (in particular) cannot yet be provided.

All the effects observed agree perfectly with those seen and reported recently in different Ti-, Al-, and Nb-bearing materials in helium of typical high temperature reactor composition; in particular there is agreement on the total oxygen consumption and on the elemental distribution in the oxide layers and in the intergranular oxide penetrating the grain boundaries.

REFERENCES

1. A. F. Weinberg and J. M. Scoffing (1962). USAEC Report GA-2998.
2. A. B. Knutsen, J. F. G. Conde and K. Peine (1969). *Brit. Corr. J.*, **4**, 94.
3. H. G. A. Bates, K. Bye and J. G. Dickson (1972). Dragon Project Tech. Note DPTN/235.
4. R. A. U. Huddle (1971). Proc. BNES Conf. on Effects of Environment on Mater. Prop. in Nuclear Systems, London (July).
5. L. Berry and H. Willermoz (1972). Proc. Conf. on Components Design in High Temperature Reactors using Helium as a Coolant, vol. 2, London (May), pp. 41–54.
6. R. A. U. Huddle and H. G. A. Bates (1973). 'Selection of Materials for the Direct Cycle High Temperature Reactor', Proc. Reactor Meeting, Deutsches Atomforum/KTG Karlsruhe.
7. H. Willermoz, L. Herry, J. Dixmier and P. Olivier (1973). 'Comportement de divers aciers et alliages dans l'helium et ses impuretés aux températures élevées', Proc. 5ème Congrès de la Fédération Européenne de Corrosion, Paris (September).
8. C. Tyzack *et al.* (1971). Proc. Conf. on Component Design in High Temperature Reactors using Helium as a Coolant, vol. 2, London (May), pp. 1–39.
9. J. Dixmier *et al.* (1975). Proc. IAEA Symposium on Gas-cooled Reactor with Emphasis on Advanced Systems, vol. I, IAEA-SM-200/61 (July), pp. 363–77.
10. M. J. Bennett and M. R. Houlton (1977). AERE-R 8235.

DISCUSSION

R. Hales (BNL, Berkeley, UK): We were involved in the materials assessment for the Dragon project and found that high vacuum techniques could be used instead of high pressure helium; impurity elements could be introduced much more easily. We agree with these authors that Nb has superior internal oxidation resistance to titanium and aluminum. In alloy 800 there is a tendency to internal attack which increases with increasing time and temperature but decreases with decreasing pO_2. Do the authors see similar effects?

Reply: We agree on the influence of time and temperature. When we changed the pH_2/pH_2O ratio (50 → 100), the depth of internal attack of alloy 800 was about constant.

M. Dixmier (Commissariat à l'Energie Atomique, France): Could you please let me know what materials are used for the capillary tubes set up in front of and behind the Cu/CuO_2 furnace, and what is their size (length, inner diameter, outer diameter)?

Reply: The capillary tube (od 2 mm; id 0.6 mm) was made of pure nickel (99.2 %). A length of 10 cm each was twisted to form a spiral of 10 mm diameter.

3

Influence of Zirconium as an Alloying Addition on the Corrosion Behaviour of a Ni/15Cr Alloy in an Oxygen/Sulphur Dioxide Atmosphere at 850°C

K. N. STRAFFORD and P. J. HUNT

Newcastle upon Tyne Polytechnic, UK

ABSTRACT

The corrosion behaviour of various ternary Ni/15w/o Cr/Zr-type alloys, with Zr contents in the approximate range 0.5–10w/o, has been determined in an oxygen/sulphur dioxide atmosphere (volume ratio 4:1) at 850°C, and compared with the performance of a Ni/15Cr binary control alloy, in terms of overall kinetics, and the patterns of scaling and of subsurface degradation. It has been established that ternary alloys containing less than about 5w/o Zr corrode more slowly than the binary materials, and the optimum level of Zr content for maximum reduction in overall weight gain appears to occur at about 1w/o, as reflected in a twenty-fold decrease in the measured parabolic rate constant. The alloy containing ~10w/o Zr, in contrast, underwent more overall wastage than the control material.

Such a pattern in behaviour relating to Zr content was also mirrored in the differences between the average thicknesses of external oxide scales from alloy to alloy; the patterns of subsurface degradation (principally accounted for by subsurface oxidation), although more complex, were also evidently governed by the Zr content in the alloy and in particular the amount and distribution of the intermetallic phase Ni_5Zr. The thinner scale formed on the Ni/15Cr/1Zr alloy appeared to be single-layered, chromium rich and essentially devoid of nickel, in contrast to the thicker duplex scale, made up of nickel oxide and a nickel–chromium oxide, produced on the control alloy.

While this observation is undoubtedly a major factor associated with the decrease in overall weight gain of the ternary relative to the binary material, the change in the type of subsurface degradation observed with the ternary alloy, relative to the control, is also significant. The factors which may be responsible for the observed modification of corrosion behaviour are considered.

INTRODUCTION

In previous studies the isothermal corrosion behaviour of various Ni/15w/o Cr/X-type ternary alloys was broadly characterised in pure oxygen (1), in a hydrogen–hydrogen sulphide atmosphere (containing about 4% H_2S) (2), and in an oxygen–sulphur dioxide environment (volume ratio 4:1) (3) at temperatures of 850 or 900 °C, and compared with the corresponding performance of a Ni/15Cr control alloy. Here X is a refractory sulphide-forming element and refers specifically to the elements La, Ce, Y, Sm, Gd, U, Th, V or Zr at concentration levels in the range of about 0.5–5w/o. These evaluation studies (4) established that certain of these reactive elements, in particular Zr, were beneficial in decreasing the rates of corrosion in the atmospheres.

The observations made in the bioxidant oxygen–sulphur dioxide environment were especially interesting in that addition of Gd, U, V or Zr at levels up to 5w/o altered beneficially the patterns of scaling and of subsurface degradation, relative to the control alloy. Additions of Zr were particularly effective although there was some indication that the effects were dependent on the Zr content of the alloy (3).

The objects of the present chapter are two-fold: first, to describe the results of further and more detailed corrosion studies which have been carried out on a number of ternary Ni/15Cr/Zr alloys, with Zr contents in the approximate range 0.5–10w/o, and on a Ni/15Cr control alloy in an oxygen–sulphur dioxide environment at 850 °C, and secondly to consider some of the factors which may be responsible for the observed modification of corrosion behaviour associated with the Zr additions.

EXPERIMENTAL

The alloys were prepared by vacuum induction melting high purity elements. Five ternary nickel-based chromium-containing alloys were prepared of differing Zr contents. A binary Ni–Cr alloy was also produced: the chromium content of all of the alloys was standardised at about 15w/o. The actual compositions (w/o) of the alloys are shown in Table 1.

Thin discs (about 2.10–2.20 mm thick) were sliced from each ingot using a diamond-edged cutting wheel. Specimens of approximate dimensions 9.50 × 3.50 × 2.00 mm were then cut/ground from these discs: such specimens had a surface area of about 1.20 cm² and weighed about 500 mg. A small hole (1.5 mm in diameter) was drilled in a corner of each coupon,

TABLE 1

	Control alloy	Ternary alloy				
		1	2[a]	3[a]	4	5[a]
Cr	15.19	14.87	15.00	15.08	14.05	14.88
Zr	—	0.68	0.89	4.92	5.60	9.25

[a] Alloys used to determine overall kinetics of corrosion (see Fig. 1).

and samples were then annealed for 3 h at 950 °C in vacuum (about 0.02 torr residual air pressure) and furnace cooled over a period of about 2–3 h. The surfaces of samples were then ground down successively from a coarse SiC paper (180 grade) to a final finishing grade of 600.

Specimens were suspended above a vitreous silica boat with platinum wire. Suspended specimens (and the boat) were then degreased ultrasonically in methanol and dried in a hot-air blast. The loaded boat was inserted and sealed into a horizontal tube furnace and the oxygen/sulphur dioxide gas mixture (dried and premixed, volume ratio 4:1) passed through the tube at a high rate of about 1360 ml/h for 1.5 h to flush out the residual air. At this stage the coupons were at room temperature. The gas rate was then reduced to a lower level of 170 ml/h, which was maintained for the duration of an experiment. The furnace (already at 850 °C) was then quickly rolled (motor driven) along the length of the tube to a fixed position and the experiment commenced: thus the alloy samples were brought rapidly to temperature in the gas stream. Specimens were exposed for individual periods of exposure up to a maximum of 100 h. At the completion of an experiment the furnace was driven back to the other end of the reaction tube, and samples were thus rapidly cooled in a standard manner in the O_2–SO_2 gas stream.

The samples were weighed before and after exposure to the gas mixture. The accuracy of measurement was ± 0.04 mg, or about ± 0.03 mg/cm^2 for samples with a surface area of about 1.2 cm^2.

Corroded samples of the alloys were examined metallographically in section, and the distribution of elements (Ni, Cr, Zr and S) through the scales and within the substrates assessed using a Cambridge SEM Mk 2 instrument fitted with an energy dispersive X-ray analyser system attachment. Mounting and polishing techniques have been described elsewhere (5). Because of the thin nature of the oxide scales a taper sectioning technique was adopted (6): this enabled scale thicknesses to be determined with considerable precision.

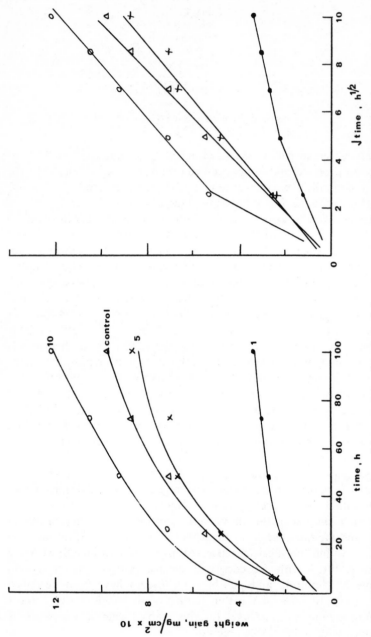

FIG. 1. Weight gain as a function of time, and as a function of the square root of time, observed during the corrosion of a Ni/15w/oCr (control alloy) and Ni/15Cr/Zr alloys with Zr contents of about 1, 5 and 10 w/o, in an oxygen–sulphur dioxide environment at 850 °C

RESULTS AND DISCUSSION

The overall weight gains observed with the binary control alloy, and the ternary alloys containing about 1, 5 and 10% of zirconium, during isothermal exposure to the oxygen/sulphur dioxide atmosphere at 850 °C, have been plotted as a function of time and as a function of the square root of time, in Fig. 1. It is apparent that the kinetics are protective for all the materials; the precise form of the kinetics is not unequivocally clear on the basis of the discontinuous weight gain/time measurements, particularly in the early stages of reaction, but it can be assumed that the kinetics are approximately parabolic for all the alloys.

Calculated parabolic rate constants (k_p) are given in Table 2. From this it can be noted that the rate constants for the control alloy, and the ternary alloys containing about 5 and 10% Zr, are essentially similar at about 2.5×10^{-12} g^2/cm^4/s, whereas the alloy containing about 1% Zr corrodes at a smaller rate, a value of 0.13×10^{-12} being obtained which is about 20 times smaller than the k_p value for the control alloy.

Because of the somewhat approximate values of these rate constants a comparison with other published data should perhaps be treated with some reserve. Nevertheless it is of interest to note that a parabolic rate constant of about 0.5×10^{-12} g^2/cm^4/s may be inferred from the literature (7) for chromium oxidation in pure oxygen at 850 °C, i.e. for Cr_2O_3 formation, while the value for pure nickel may be taken to be about 10^{-11} (8). Thus it seems reasonable to assume that the scaling kinetics observed in the present studies with an alloy containing about 1% Zr essentially reflect the growth of a predominantly 'Cr_2O_3' scale. The actual nature and composition of the scale formed on the alloy is considered below.

It is also of interest to compare the overall weight gains from alloy to alloy after a prolonged period of corrosion. In Table 2 are quoted observed weight gains after 72 h exposure. It is now clear that the largest amount of corrosion is experienced by the alloy containing 9.25% Zr, although the weight gain exhibited by an alloy containing 5.6% Zr is only slightly lower. In fact both of these weight gains are higher than that characteristic of the control. Clearly the alloys containing 0.68 and 0.89% Zr exhibit the lowest overall weight gains.

It is useful to compare graphically the overall weight gains noted after this period of exposure at different levels of Zr content—this has been done in Fig. 2. It is emphasized that the drawing of a continuous line between the discrete points in Fig. 2 is not meant to imply a functional relationship, but rather to emphasize trends. Thus there appears to be a minimum corrosion

K. N. Strafford and P. J. Hunt

TABLE 2

Corrosion parameters appertaining to a Ni/15Cr alloy and various Ni-Cr-Zr alloys containing ~0.5-10% Zr when exposed to an O_2/SO_2 environment at 850°C (Cr content in Ni-Cr-Zr alloys: 14.9 ± 0.15 w/o)

Alloy composition (w/o)		Parabolic rate constant $(g^2/cm^4/s)$	Weight gains after 72 h (mg/cm^2)	Scale thicknesses after 72 h (μm)[a]			Depths of subsurface degradation after 72 h (μm)[b]		
Nominal	Actual			Min	Average	Max	Min	Average	Max
Ni-15Cr	15.2	2.86×10^{-12}	0.86	2.3	4.7	14.2	17	34	47
~1	0.68Zr	—	0.35		—			—	
	0.89Zr	0.13×10^{-12}	0.32	1.6	2.5	4.0	16	32	56
~5	4.92Zr	2.34×10^{-12}	0.75	0.8	3.1	6.1	18	26	32
	5.60Zr	—	1.03		—			—	
~10	9.25Zr	2.25×10^{-12}	1.07	1.6	4.0	7.2	16	31	41

[a] Accuracy ± 0.2 μm (taper sections).
[b] Accuracy ± 2μm.

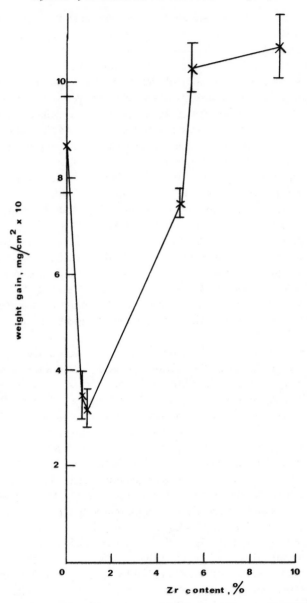

FIG. 2.   Weight gains observed after the exposure of various Ni/15Cr/Zr alloys, with Zr contents in the range 0–9.25 w/o, to an oxygen/sulphur dioxide environment at 850 °C for 72 h

rate in the alloy system associated with a zirconium content of about 1 %. At Zr contents either greater or less than this minimum the extent of corrosion increases.

The overall kinetics in fact represent a summation of total weight gain accounted for both by external oxide scale formation and also a degree of subsurface degradation in the form of particles/stringers of sulphide and/or oxide. The average thicknesses of the scales and the depths of subsurface degradation after a 72 h exposure period are given in Table 2. It is evident that while there are little differences between the average depths of internal degradation from alloy to alloy, there are significant differences in scale thickness. In particular it may be noted that the scale formed on the Ni/15Cr/0.89Zr alloy is about half the thickness of that formed on the control alloy. In fact the scales formed on all of the Zr-bearing alloys are all thinner than that formed on the control alloy.

Bearing these data in mind it seems clear that the differences in the overall amounts of corrosion from alloy to alloy, and in particular between the control alloy and the alloy containing about 1 % Zr, are principally associated with a modification to the pattern of scaling rather than varying depths of internal degradation. Thus it must be assumed that the Zr is somehow responsible for a significant modification to scaling behaviour.

In this context it is appropriate to consider the structure and composition of the scales formed on the various alloys. Scales were examined microscopically and the distribution of Ni, Cr, Zr and S assessed qualitatively using an energy dispersive X-ray analyser system fitted to a scanning electron microscope. The main features to emerge from these studies have been summarized in Figs. 3 and 4. The oxide scale formed on the Ni/15Cr control alloy was duplex, consisting of an outer layer which was nickel-rich (presumably NiO), surmounting an inner layer of a nickel–chromium oxide, which was chromium-rich, its composition probably approximating to that of the spinel $NiCr_2O_4$. In complete contrast the scale formed on the alloy containing about 1 % Zr which, as we have seen, exhibited both the lowest overall weight gain and the smallest scale thickness, was apparently single-layered and chromium rich, and in particular contained no nickel. Thus the presence of Zr in the alloy has apparently had the effect of inhibiting the formation of NiO as an outer layer, and furthermore although the exact content of nickel in the chromium-rich oxide could not be determined it was clearly negligible. Significantly a small amount of zirconium was observed in the oxide formed on the alloy containing about 1 % Zr, while the presence of this element was

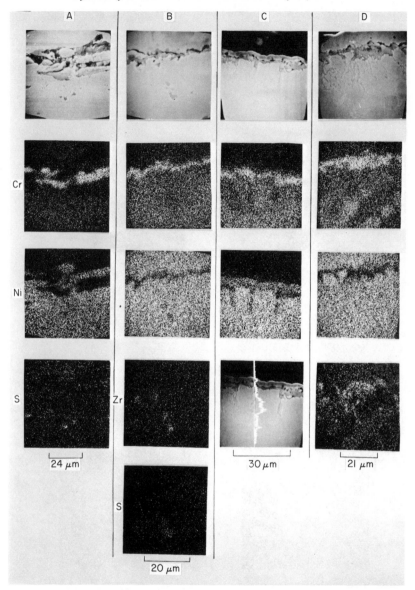

FIG. 3. Morphological features of corrosion relating to (A) the binary control alloy, (B) the Ni/15Cr/1Zr alloy, (C) the Ni/15Cr/5Zr alloy, and (D) the Ni/15Cr/10Zr alloy—electron and X-ray images, as indicated. Exposure conditions, 72 h at 850 °C

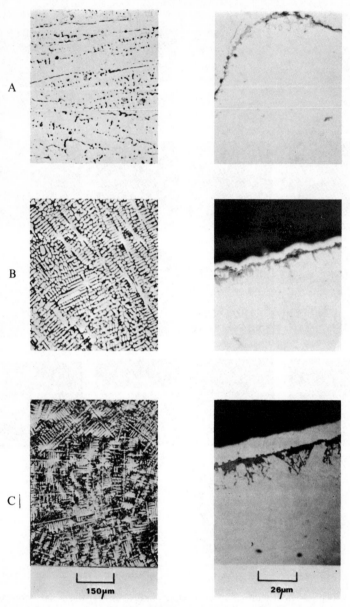

FIG. 4. Microstructures of the Ni/15Cr/Zr alloys, and morphological features of the associated corrosion products, 72 h exposure at 850 °C. Alloy A contains ~1 % Zr, B ~ 5 % Zr and C ~ 10 % Zr

readily detected in scales formed on the alloys containing about 5 and 10 % Zr.

It seems therefore that a degree of selective attack of Zr has occurred in all the Zr-bearing alloys, a greater amount being oxidized the larger the Zr content of the alloy. In fact an examination of appropriate thermodynamic data, in particular a comparison of the free energies of formation of Cr_2O_3, ZrO_2 and NiO, would suggest such a pattern, although there is a need to attempt to calculate the level of Zr at which selective attack of this element might be anticipated.

The most interesting and significant observation however is the absence of NiO in the Zr-bearing alloys. There is clearly a need to examine this feature in detail, but it would appear that the nucleation of NiO and/or its growth in the early stages of corrosion is essentially prevented in the presence of zirconium. This may well be associated with the encouragement of a higher nucleation rate of Cr_2O_3 in a manner analogous to the ideas put forward by Stringer *et al.* (9) and others (10, 11) concerning the possible role of certain reactive metals, e.g. La, Ce or Y (and their oxides) in chromium-containing heat-resisting alloys. It is probably reasonable to assume that the subsequent growth of Cr_2O_3 will be more rapid in the presence of Zr (in solid solution) on account of the anticipated increase in vacancy concentration brought about by a Wagner-type doping mechanism. However it is also worth noting that the (parabolic) growth rate of ZrO_2 is about 10^{-9} g^2/cm^4/s at 850 °C (12), i.e. about four orders of magnitude faster than the growth rate of Cr_2O_3 and significantly two orders faster than the growth rate of NiO. In the oxidation models for Ni–Cr alloys proposed by Wood (13), an essential feature is the overgrowth of the Cr_2O_3 by NiO on account of this difference in growth rate, which leads to the production of an outer layer of NiO. Thus it seems reasonable to assume that the overgrowth of nucleated NiO by ZrO_2 (and doped Cr_2O_3) may be at least partially responsible for the absence of NiO in the scales formed on these Ni–Cr–Zr alloys. However there is an obvious need to examine in detail the early stages of corrosion, a study which is currently being undertaken in these laboratories.

Although the average depths of internal degradation (see Table 2) were about the same from alloy to alloy ($\sim 30 \, \mu$m), irrespective of Zr content, there were subtle and important differences in the composition and morphology of the corrosion products, particularly in the population densities of oxide or sulphide particles or stringers. Thus internal degradation in the Ni/15Cr control alloy was associated primarily with the formation of rounded 'CrS' particles of an average density of distribution, occupying some 4.5 % by volume of the substrate material, as determined by a point-counting method. In contrast, in the alloys containing about 5 and 10 % Zr, no sulphides were detected in the zone of degradation: rather, oxide in the form of stringers (containing both Cr and Zr) was observed with a higher population density—about 5.4 and 8.2 % by volume, respectively. The volume of corrosion product and related particle density, observed with the alloy containing about 1 % Zr was smallest at about 1.0 %, and here both zirconium oxide stringers, and sulphide particles of zirconium and chromium, were detected.

These observations may indicate that the scale formed on the Ni/15Cr/1Zr alloy does not permit as high an inward flux of sulphur as does the scale formed on the binary control—that is, considering the diffusion to be regular—i.e. occurring via a vacancy mechanism, and ignoring the possibility of mechanically imperfect (cracked or porous) scales—the diffusion coefficient and/or solubility of sulphur in the 'Cr_2O_3' scale formed on the Ni–Cr–Zr alloy is/are smaller than in the scale formed on the binary alloy. The possible role of the presumably different structures of the oxide scales in respect of grain size and type (equiaxed or columnar) needs clarification in this context.

An important additional factor probably relates to the different fluxes of sulphur in the alloy substrates once the sulphur has penetrated the external 'protective' scales thereby allowing the formation of internal sulphides deep in the alloy substrates. In this context Strafford and Harrison (2), in studying the sulphidation behaviour of various Ni/15Cr/X alloys, where X is one of a number of refractory sulphide-forming elements including Zr, at about the 0.5w/o level, suggested that the observed decreased depth of penetration of sulphur into the Zr-bearing substrate relative to the substrate of the control alloy might be associated with a decreased solubility of sulphur and/or diffusion coefficient. At the same time the gettering action of Zr in forming ZrS particles was anticipated and noted. It seems that such factors could also be operative in the present situation.

It seems clear that the relatively high corrosion rates experienced by the alloys containing ~5 and particularly 10 % Zr are associated with the increasing amounts of the intermetallic phase Ni_5Zr which is encountered in the Ni–Zr alloy system with Zr contents less than 12 % Zr. Examination (Fig. 4) of the cast microstructures and the corresponding corroded sections of alloys indicated clearly the susceptibility of the Ni_5Zr phase to undergo selective oxidation, which principally occurs in grain boundary regions. The correlation between intergranular oxidation and intermetallic particle distribution is particularly evident with the Ni/15Cr/9.25Zr alloy; in fact, comparison of the average densities of subsurface corrosion products noted earlier, and the amount of Ni_5Zr calculated to be present in the alloys on the basis of the (binary) phase diagram, reveals a direct and linear correlation. It is also clear that the relatively small and well-distributed amount of the Ni_5Zr phase in alloys containing about 1 % Zr will provide a good keying-on effect for the scale, following its oxidation, to form very small and thin stringers in the substrate. This keying-on might indirectly control the degree of imperfection (porosity) in the scale, thus also contributing to a decreased penetration by sulphur, compared with the penetration experienced by the control.

In conclusion it has been established that additions of zirconium to a nickel-15w/o chromium alloy significantly modify its corrosion behaviour when exposed to an oxygen–sulphur dioxide atmosphere at 850 °C. Specifically an addition of about 1w/o Zr considerably improves the corrosion performance with four principal attendant features: (1) a substantial decrease in the parabolic rate constant, and in overall weight gain, after a prolonged period of exposure; (2) a significant decrease in external oxide scale thickness; (3) the production of an oxide scale with totally different nature and composition; and (4) a changed mode of subsurface degradation, both in terms of the degree (depth) and intensity (particle density) of attack, and the composition of the internal corrosion product(s).

It seems that this modification to corrosion behaviour could be attributable to several interrelated factors, and currently studies are being carried out in these laboratories in an attempt to elucidate the mechanism(s) involved.

ACKNOWLEDGEMENT

This work has been carried out with the support of the Procurement Executive, Ministry of Defence.

REFERENCES

1. K. N. Strafford and J. M. Harrison (1976). *Oxid. Met.*, **10**, 347.
2. K. N. Strafford and J. M. Harrison (1974). Proc. Electrochem. Soc. Conf., Toronto, p. 464.
3. K. N. Strafford, P. J. Hunt and G. R. Winstanley (1976). 'Modifications to the

Pattern of Isothermal Corrosion Observed with a Ni/15Cr Alloy in a O_2/SO_2 Atmosphere Associated with Additions of Certain Reactive Elements', paper presented at Electrochemical Soc. Mg, Las Vegas, Nevada (October).
4. K. N. Strafford, P. J. Hunt, G. R. Winstanley and J. M. Harrison (1976). 'Evaluation Studies Concerning the Potential of Certain Group II–V Elements as Additions to Nickel-based Alloys for the Inhibition of Hot Corrosion', Proc. 3rd US/UK Navy Conf. on Gas Turbine Materials in a Marine Environment, Bath University (September).
5. K. N. Strafford and A. F. Hampton (1970). *J. Less Comm. Met.*, **21**, 305.
6. F. Weinberg, Ed. (1970). *Tools and Techniques in Physical Metallurgy*, Marcel Dekker, New York.
7. W. C. Hagel (1963). *Trans. Am. Soc. Met.*, **56**, 583.
8. C. S. Giggins and F. S. Pettit (1969). *Trans. Am. Inst. Min. Engrs.*, **245**, 2495.
9. J. Stringer, B. A. Wilcox and R. I. Jaffee (1972). *Oxid. Met.*, **5**, 11.
10. I. G. Wright, B. A. Wilcox and R. I. Jaffee (1972). Final Report Manual Air Systems Command Contract N00019-72-C-0190 (December).
11. G. C. Wood (1971). *Werkstoffe Korros.*, **22**, 491.
12. K. H. Akram and W. W. Smeltzer (1962). *Can. Met. Qly.*, **1**, 41.
13. G. C. Wood (1970). *Oxid. Met.*, **2**, 11.

DISCUSSION

A. R. Nicoll (BBC Forschungslabor, Heidelberg, Germany): Is the scale on the 1 % Zr variation mechanically stable?

Reply: The protective properties of the scale formed on the Ni/15Cr/1Zr alloy appear to be excellent. Its ability to resist penetration by sulphur seems distinctly better than that associated with the scale formed on the binary Ni/15Cr alloy. The corrosion behaviour of the alloy and mechanical stability of the scale is good under conditions of thermal cycling.

E. Erdoes (Sulzer AG, Winterthur, Switzerland): Do the elements of the IVth group (Ti, Zr, Hf) improve, in general, the oxidation properties?

Reply: Small amounts ($\sim$ 1w/o) of zirconium certainly improve the oxidation behaviour of a Ni/15Cr alloy—the effects of the other Group IV elements, titanium and hafnium, have not been assessed in our studies. However it is likely that such elements will be beneficial—the beneficial role of small amounts of several reactive elements from Groups II–V on oxidation behaviour has recently been reported. [K. N. Strafford and J. M. Harrison (1976). *Oxid. Met.*, **10**, 347].

H. Pflug (Robert Bosch GmbH, Stuttgart, Germany): Do you think that Zr also improves the corrosion resistance of Ni alloys with a content of 95 % Ni and small amounts of Mn and Cr (both about 2 %)?

Reply: Although the detailed mechanism whereby Zr is effective in reducing the corrosion rate of a Ni/15Cr alloy in oxidizing environments is not clear, and indeed is presently under study at Newcastle, it appears that a key factor is the preferential formation of a very protective Cr_2O_3-rich scale in the early stages of corrosion. Zirconium apparently greatly encourages the nucleation and growth of Cr_2O_3 in these early stages, and it appears that additions of Zr can be empirically equated with Cr levels in the alloy. Thus at the 15 % Cr level, experience suggests that the addition of 1 % Zr is equivalent (in its effect in reducing the rate of corrosion) to the presence of ~ 5 % Cr, i.e. the performance of a Ni/15Cr/1Zr alloy $\equiv$ Ni/20Cr.

It may therefore be that Zr will also be effective in reducing the corrosion rate of a 95 % Ni alloy containing ~ 5 % Cr, but we have not examined the behaviour of such an alloy. Possibly the 'effective' Cr level even in the presence of Zr will be too low.

J. Stringer (University of Liverpool): If I can add to Dr Strafford's answer to Dr Erdoes: both Dr Strafford and ourselves have found that the maximum effect of these reactive additions (including the Group IV metals) is at relatively low concentrations: typically 1 % or less—much less than the Ti content in IN 738. By the time the reactive metal content has reached this level, the beneficial effect has largely disappeared. I think, therefore, that we cannot conclude that the good behaviour of IN 738 and similar alloys in hot corrosion can be attributed to the sort of effects discussed by Dr Strafford.

Reply: Further to the comments of Professor Stringer, I think it is very difficult to be certain, at the present time, of the mechanism(s) by which the beneficial role(s) of Ti, Zr, etc., are established, whether under 'hot corrosion' conditions or oxidation, or indeed sulphidation [K. N. Strafford and J. M. Harrison (1975). Proc. Amer. Electrochem. Soc. Mg, Toronto, p. 464]. As noted elsewhere in these discussions we are currently examining in detail the behaviour of Ni–Cr–Zr alloys in an O_2/SO_2 atmosphere to try to ascertain the mechanism. Here it appears that the oxide composition (and possibly its structure, grain size, etc.) is crucial, but it may well be that under 'hot corrosion' conditions, especially in the presence of a molten deposit, different criteria may become more important. There is clearly a need to assess behaviour of alloys containing reactive additions in more detail, especially in the early stages of corrosion before steady state scaling has been achieved.

J. C. Colson (Laboratoire de Recherches sur la Réactivité des Solides, Dijon, France): Could you tell us where the sulphides formed are located?

Reply: In the Ni/15%Cr control alloy sulphides were located in the alloy beyond the scale/alloy interface. As noted in the paper, an average depth of internal attack after 72 h exposure was $\sim 34 \mu m$ and the particles were grouped in clusters. No sulphides were detected in alloys containing 5 and 10% Zr. In the alloy bearing 1% Zr it was clear that a degree of selective sulphidation (ZrS formation) of the intermetallics had taken place, and a small amount of CrS had formed, in a random manner, within the substrate.

J. C. Colson: What is your proposed mechanism for sulphur transfer to obtain sulphides at the alloy–scale interface?

Reply: It is likely that sulphur migrates too slowly through both NiO and Cr_2O_3 to allow significant penetration to the scale/alloy interface unless short-circuit paths exist such as pores or subgrain boundaries. If such short-circuit paths do exist, then sulphur can migrate to, and within, the alloy where the oxygen activity is sufficiently low to allow sulphide formation. It has been reported [C. B. Alcock, M. G. Hocking and S. Zader (1969). *Corros. Sci.*, **9**, 111] that sulphur may create cation vacancies in NiO, and it is interesting to note in the present work that it is where an exterior NiO scale is formed that the most sulphide formation is seen (Ni/15Cr control alloy).

In the case of a chromic oxide-rich scale (as formed on 1, 5 and 10% Zr alloys) it is possible that Zr^{4+} ions, which are considerably larger than Cr^{3+}, and/or ZrO_2 particles, block short circuit diffusion along grain boundaries, and hence reduce the tendency for internal sulphidation to take place.

4

Effect of Yttrium and Hafnium Additions on the Oxidation Behaviour of the System Co–Cr–Al

D. P. WHITTLE, I. M. ALLAM and J. STRINGER

University of Liverpool, UK

ABSTRACT

The effect of small amounts of hafnium or yttrium additions on the oxidation behaviour of Co–Cr–Al alloys, which form Al_2O_3 scales, has been studied. Oxidation kinetics in dry air were measured over the temperature range 1000–1200 °C, and the morphological developments of the scales studied in detail. The main beneficial effects were (1) a reduction in scale thickening rate; (2) a reduction in the initial transient stage of oxidation; and (3) a remarkable improvement in scale adherence, particularly under thermal cycling conditions. Preferential oxidation of Hf-rich and, to a lesser extent, Y-rich phases results in the formation of numerous oxide protrusions penetrating into the substrate alloy and these are largely responsible for the improvement in scale adherence. It appears that the size and distribution of the oxide protrusions play a decisive role in maintaining scale adherence over long exposure periods. In addition, the absence of voids at the alloy/scale interface in alloys containing the active metal additions, and their reappearance after long-term oxidation tests, suggest that they are Kirkendall voids and support the validity of a vacancy sink mechanism. Finally, it is shown that hafnium additions (> 0.3w/o) are more beneficial for scale adherence than corresponding yttrium additions, and this is probably related to the more uniform distribution of hafnium in the alloy matrix which results in an optimum size and distribution of the oxide pegs at the alloy/scale interface.

INTRODUCTION

In developing oxidation-resistant materials for service at elevated temperatures, two requirements must be satisfied: diffusion through the oxide scales must occur at the slowest possible rate, and the oxide scale must resist spallation, particularly when subjected to growth and thermal stresses. Al_2O_3 adequately fulfils the former requirement, but its adhesion to the substrate is poor. Attempts to improve the adhesion have largely centred on the so-called 'rare-earth effect', Y being the particular element used to improve the adherence of Al_2O_3 scales on M–Cr–Al–Y-type alloys and coatings.

A number of mechanistic models have tentatively been proposed to account for the influence of these additions in promoting scale adherence. These include: (a) mechanical keying of the oxide caused by selective oxidation of the active element at the alloy/scale interface, giving rise to protrusions or oxide pegs (1–4); (b) provision of sites for vacancy condensation at the internal particles of alloying element oxides, thereby eliminating interfacial porosity (5); (c) formation of an interlayer to act as a graded seal between oxide and substrate (6); and (d) modification of the oxide scale plasticity allowing accommodation of the growth and thermal stresses (7).

None of these mechanisms can be considered generally valid and the present study was initiated to examine in more detail the oxide/substrate interface and to establish the critical parameters. In addition, Y additions to alloys of this type have a number of disadvantages: the Y is usually present as an intermetallic compound, at the grain boundaries, and this can result in hot-working problems. Thus, a further objective of the present investigation is to see whether another reactive metal which does not form an intermetallic compound can provide similar beneficial effects.

EXPERIMENTAL

Alloys in the Co–Cr–Al system which form Al_2O_3 scales (8) were studied, Co/10Cr/11Al being the basis composition. The alloys also contained between 0 and 1.5 wt % Y or Hf. They were prepared by vacuum induction melting from 99.9 % pure starting materials, and were vacuum cast into 25 mm square section ingots. Samples having approximately 2 cm² surface area were cut from the as-cast blocks and their surfaces prepared by grinding on SiC metallographic polishing papers to 600 grit.

Isothermal oxidation tests were conducted using a conventional Sartorius electronic recording balance. The furnace at reaction temperature was raised around the mullite reaction vessel to commence the experimental run, and this was terminated by lowering the hot furnace.

Cyclic exposures were conducted in a horizontal tube furnace: samples were exposed in small recrystallized alumina crucibles for 20 h, after which they were removed from the furnace and allowed to cool for 2 h. Weight changes were determined at room temperature at the end of each cycle: samples were not removed from the crucibles for weighing and the data presented later include total weight gains, spalled plus adherent oxide.

RESULTS

Alloy Structure
The Co–Cr–Al alloys consisted of a dark, dendritic phase (β'–CoAl) in a lighter matrix (Co-rich α-phase). Alloys containing Y have a third phase, the intermetallic yttride Co_3Y (9), localized at alloy grain boundaries. Disappearance of this phase at Y concentrations below 0.1 wt % suggests that the solubility limit of Y lies somewhere below 0.1 wt %. No intermetallic phases were observed in the Hf-containing alloys, suggesting a solubility of at least 1.5 %. Addition of either element to the alloys reduced the grain size by approximately 50 %.

Isothermal Oxidation Kinetics
Figure 1 shows the oxidation kinetics at 1100 °C for the Co/10Cr/11Al/Y alloys. After an initial transient period, an external scale of α-Al_2O_3 is formed on all the alloys with an appropriate reduction in oxidation rate. Increase in alloy Y-content resulted in shorter transient periods, and only small amounts of spinel formed before the continuous Al_2O_3 layer was established. Greater amounts of spinel were formed on alloys whose compositions were closer to the borderline between regions IV and II in the Co–Cr–Al oxide map (8).

Giggins and Pettit (10) described the kinetics of oxidation of Co–Cr–Al, Ni–Cr–Al, Co–Cr–Al–Y and Ni–Cr–Al–Y alloys in terms of a parabolic rate law. This is not possible, even approximately, with the present results and log/log plots of weight change vs. time for the Co/10Cr/11Al/0.3Y alloy have slopes of 0.21, 0.24 and 0.42 at 1000, 1100 and 1200 °C, respectively.

Giggins and Pettit (10) also suggested that the presence of Y did not have any significant effect on the growth rate of the Al_2O_3 scale. Furthermore, their data are intermediate between those of the Co/10Cr/11Al alloy, with and without the Y-addition in the present study. However, the growth rate of Al_2O_3 does appear to be dependent to a limited extent on the alloy composition: Giggins and Pettit's alloys contained 25Cr/6Al/1Y.

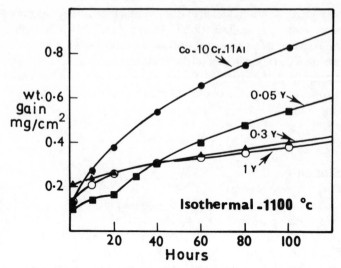

Fig. 1.   Isothermal oxidation kinetics at 1100 °C of Co/10Cr/11Al containing 0,
0.05, 0.3 and 1.0 wt % Y

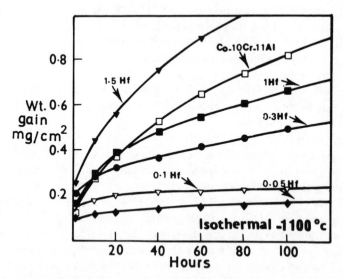

Fig. 2.   Isothermal oxidation kinetics at 1100 °C of Co/10Cr/11Al containing 0,
0.05, 0.1, 0.3, 1.0 and 1.5 wt % Hf

Figure 2 shows the influence of Hf-additions on the isothermal oxidation kinetics at 1100 °C of Co/10Cr/11Al. Hf also appears to reduce the initial transient stage of oxidation, as well as decreasing the overall rate. The minimum rate is associated with the lowest Hf additions, 0.05 and 0.1 wt %. Increasing the Hf content causes increased rates, although up to 1 % Hf additions rates are still less than the Hf-free alloy: the 1.5 % Hf alloy oxidizes at a faster rate. This behaviour is in contrast to that of the Y

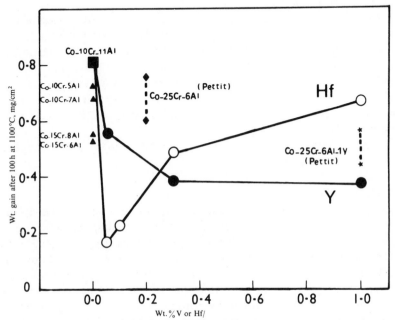

FIG. 3.   Weight gain after 100 h exposure at 1100 °C as a function of alloy Hf or Y content: Co/10Cr/11Al alloys

additions, where the greater the Y content (up to 1 wt %) the lower the overall weight gain.

Because the rate curves are not parabolic, it is not possible to compare the effects of the additions in terms of a well-defined rate constant. Figure 3 compares the weight gain after 100 h exposure at 1100 °C. On the active-element-free alloys, the weight gain ranges from 0.53 to 0.82 mg/cm² and there is no real systematic variation with alloy composition. Y additions of 0.3 % or more produce a marginal reduction in the weight gain in comparison with the best Y-free alloy. However, the 0.05 and 0.1 % Hf alloys show substantial improvements.

Thermal Cycling

As indicated earlier, the major effect of the active element additions is in promoting scale/substrate adherence, and this is more critical during thermal cycling tests than isothermal exposure. A 1 wt% Y addition to Co/10Cr/11Al reduces the overall weight gain after nine 20 h cycles at 1200 °C from 65 mg/cm^2 to 4.27 mg/cm^2. 0.1 and 0.3 wt% Y additions are almost as effective, but the 0.05% Y-containing alloy behaves little differently from the ternary.

With the Hf-containing alloys a minimum of 0.3% Hf is necessary to maintain good scale adherence: increase in the Hf content above this level is detrimental. Alloys containing 0.05 and 0.1% Hf, which showed the slowest overall growth rates in isothermal tests, behaved little differently to the undoped alloys during thermal cycling. Heavy scaling with these alloys was largely confined to specimen corners and edges where CoO and CoAl$_2$O$_4$ had been formed following the spallation of the Al$_2$O$_3$ scale formed during the first cycle.

The best resistance to scale spallation of all the alloys is achieved with a 0.3 wt% Hf addition.

Oxide Morphology

After very short exposure times the thin scales formed on the Y- and Hf-free alloys tended always to spall from the surface on cooling. Examination of the underside of the scale revealed an Al$_2$O$_3$ grain size in the range 0.5–3 μm after 5 min at 1100 °C, increasing to 1–6 μm after 30 min. In addition, the scale had apparently a more 'open' structure after the longer oxidation time with deep interstices between the Al$_2$O$_3$ grains. The alloy surface, where the scale has spalled away, is decorated with numerous voids, many of which appear to coincide with substrate grain boundaries where scale nodules of the faster growing CoO or Cr$_2$O$_3$ have developed.

Much the same sort of behaviour in these early stages was observed with the Hf- and Y-containing alloys. There were slightly more interfacial voids in the 0.05% Hf alloy than in the 0.05% Y alloy: but in both cases more than in the undoped alloy, Co/10Cr/11Al.

Figure 4 shows the surface of the alloy (Co/15Cr/8Al) after spallation of the Al$_2$O$_3$ scale during cooling following isothermal exposure for 190 h at 1200 °C. The surface contains numerous smooth areas (about 50% of the interface area) surrounded by rough areas imprinted by oxide grains. Microprobe analysis revealed no compositional differences between the smooth and imprinted areas. The features of the smooth areas suggest they are voids formed during exposure: intimate contact between scale and alloy was only maintained at areas where the oxide imprints are observed on the alloy. However, void formation does not affect the rate of growth of the Al_2O_3; the total scale thickness above the voids is similar to that at other locations in the scale. Giggins and Pettit (10) have calculated that evaporation of Al from the alloy is well able to supply Al at the rate demanded by the growing scale.

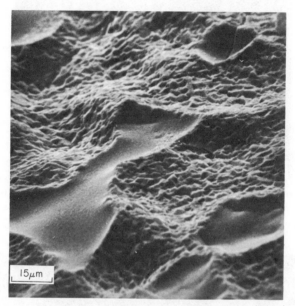

FIG. 4.   Surface of Co/15Cr/8Al following scale spallation: sample exposed for 190 h at 1200 °C

The formation of voids was observed along the interfaces of all the undoped alloys. In addition, deep substrate grain boundary valleys were observed in some locations on the Co/10Cr/11Al alloy.

Examination in cross-section of the detached Al_2O_3 scale formed on Co/15Cr/8Al shows a gradation in oxide grain size through the section (Fig. 5). The sample has been oxidized for 1000 h at 1200 °C and the total scale thickness is about 20 μm. The oxide grain size in the outer region of the scale is around 0.5 μm, increasing to about 5–10 μm at the inner surface. This gradual increase in Al_2O_3 grain size across the scale appears consistent with a scale growth model of inward oxygen diffusion, new oxide being formed at the oxide/alloy interface. Clearly, there has been no scale breakaway at

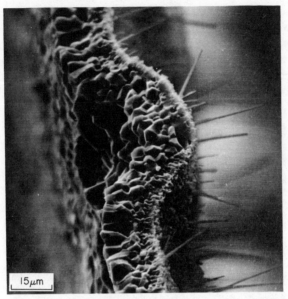

FIG. 5.   Cross-section of the detached Al_2O_3 scale formed on Co/15Cr/8Al
oxidized for 1000 h at 1200 °C

FIG. 6.   Cross-section of the scale formed on Co/10Cr/11Al/0.3Y oxidized for
130 h at 1200 °C

temperature, otherwise there would be new, small grained Al_2O_3 crystals located within the scale section. Oxide whiskers protruding outwards at the scale/gas interface can also be observed. These whiskers are approximately 0.5 μm in diameter and are also found on similar alloys containing 0.05 and 0.1 % Y but not with alloys of higher Y content. Whiskers are often associated with compressive growth stresses.

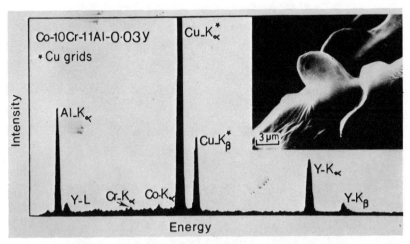

FIG. 7. Underside of the Al_2O_3 scale stripped from the alloy surface: Co/10Cr/11Al/0.3Y oxidized for 75 h at 1200 °C. The X-ray energy dispersive analysis is at the tip of one of the oxide protrusions

Multiple layers of Al_2O_3 are sometimes formed, particularly near specimen edges and corners, even during isothermal exposure.

Adding 0.05 % Y had little effect. However, 0.1 % produced a major change in the interface morphology. No smooth areas were detected, the entire alloy surface being imprinted with oxide grains. The other distinctive feature of the Y-containing alloys is the formation of oxide protrusions penetrating into the alloy. Figure 6 shows a typical cross-section. The protrusions are generally localized at the intersections of substrate grain boundaries with the alloy/scale interface. These are the sites of the Y-rich intermetallic phase in the alloy, although after oxidation the Y exists as internal oxide. However, closer scrutiny indicates the oxide pegs consist primarily of Al_2O_3: the Al_2O_3 has grown preferentially at these points encapsulating the Y-rich internal oxide particles. Energy dispersive analysis of the oxide protrusions (Fig. 7), extending from the underside of the scale, confirms the presence of Al and Y: no significant concentration of Y could be detected at other locations.

Further increase in the Y content to 0.3% produces further increases in scale adherence. In comparison to the 0.1% Y alloy, the number of oxide pegs formed at the alloy/scale interface increases: their size remains much the same.

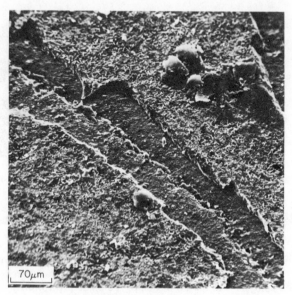

Fɪɢ. 8.   Surface of the scale formed on Co/15Cr/8Al/1Y oxidized 120 h at 1200 °C

When the Y content is increased to 1.0%, although the number of oxide protrusions is greater, their size too shows a significant increase. This situation results in a remarkable decrease in Al_2O_3 scale adherence, and frequent spalling takes place. Scale cracking, when it occurs, initiates above the large protrusions. Equally, aligned protrusions along the substrate grain boundaries can give rise to scale loss at these locations, as shown in Fig. 8. The Al_2O_3 forming the peg is often pulled out from the alloy along with the surrounding scale; this may well be related to the relatively smooth interface between peg and alloy, as was shown in Fig. 7.

Finally, after relatively long exposures, 1000 h at 1200 °C, of alloys containing lower amounts of Y, 0.3% and less, small voids appear at the alloy surface beneath the surface scale. The voids are located at the uppermost part of the boundaries between adjacent oxide grain imprints.

The voids are only of the order of 1–2 μm across and are not observed after corresponding exposures of the alloy containing 1 % Y.

An 0.05 % Hf addition to Co/10Cr/11Al is sufficient to completely eliminate the formation of smooth, voided areas at the alloy/scale interface: only single layers of Al_2O_3 are observed, which are only removed with extreme difficulty to reveal the substrate.

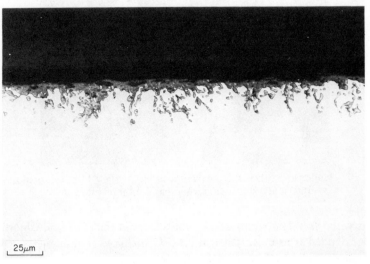

25μm

FIG. 9. Cross-section of the scale formed on Co/10Cr/7Al/1Hf oxidized for 17 × 20 h cycles at 1200 °C

Oxide protrusions at the alloy/scale interface start to become noticeable as the Hf content of the alloy is increased. Figure 9 shows the random distribution of pegs occurring at the metal/oxide interface of the Co/10Cr/7Al/1Hf alloy, in sharp contrast to the relatively poor distribution of protrusions in the Y-containing alloys. Occasionally, relatively long pegs develop, and scale loss is more prevalent above these long pegs. As indicated earlier, the oxide protrusions consist primarily of Al_2O_3, which has grown inwards encapsulating the Hf-rich oxides formed as an internal oxide precipitate.

Figure 10 shows the underside of the Al_2O_3 scale formed on Co/10Cr/11Al/1Hf in 1000 h at 1200 °C, after the alloy substrate has been dissolved away. The profusion of inwardly growing pegs is plainly visible, and these clearly provide a better key in anchoring the scale to the substrate than the relatively smooth pegs formed around the Y-rich internal oxides.

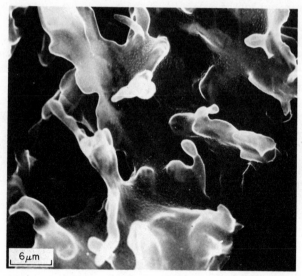

FIG. 10.   Underside of the Al_2O_3 scale formed on Co/10Cr/11Al/1Hf oxidized for
1000 h at 1200 °C following dissolution of substrate

As with the Y-containing alloys, voids form at the alloy/scale interface
after extended exposures, 1000 h at 1200 °C, on alloys containing 0.1 % Hf,
but not with alloys of higher Hf content.

DISCUSSION

The reduction in isothermal growth rate of Al_2O_3 produced by the Y and
Hf additions depends on a number of factors. Scale spallation can occur,
and while this is not extensive under isothermal conditions, re-formation of
Al_2O_3 and the development of multilayered scales can contribute
significantly to the weight gain. Differences in multilayer scale formation
are thought to be the main reason for the different overall weight gains with
the active-metal-free alloys.

The results of the thermal cycling tests clearly demonstrate that Y or Hf
reduces scale spallation, and under the relatively mild conditions of
isothermal exposure multilayer scale formation has probably been entirely
eliminated. However, it would appear that the beneficial effects of Y
(0.3 wt % and above) and Hf (<0.3 wt %) go beyond this, and actually

reduce the growth rate of the Al_2O_3, as well as preventing any scale loss. Hf also seems to be the more effective.

Neither Hf nor Y are detected to any appreciable level in the Al_2O_3 scales, although it is appreciated that very small impurity levels in Al_2O_3 could significantly alter the self-diffusion coefficients: Al_2O_3 is supposed to exhibit an extrinsic defect structure below 1650 °C (11). Trace levels of Co or Cr may also be significant. However, variation of the grain size of the Al_2O_3 may well be more significant, particularly in view of the above hypothesis that continued growth is primarily by inward short-circuit diffusion of oxygen.

The grain size varies through the scale cross-section, ranging from approximately $0.5 \mu m$ at the scale/gas interface to $5–10 \mu m$ near the oxide/alloy interface. Furthermore, this change is more marked with active-metal-containing alloys, and the reduction in Al_2O_3 growth rate could be due to an elimination of some of the short-circuit paths for inward oxygen diffusion. Such a mechanism would also be consistent with the instantaneous oxidation rate decreasing at a rate greater than that required by a parabolic rate law: as the oxidation rate declines, larger grains of Al_2O_3 are formed, resulting in fewer grain boundaries and an even further reduction in rate.

The main factor contributing to improved scale adhesion is the formation of oxide protrusions penetrating into the alloy. In contrast to earlier models, however, the protrusions or pegs are not the oxide of the reactive element, but Al_2O_3 which has grown inwards encapsulating the particles of oxide of the reactive element. The formation of these stringers makes a significant contribution to the overall weight gain and thus, under isothermal conditions, where scale spallation is not very severe the alloy with the lowest Hf content, with the fewest number of these stringers, is the most resistant. The concentration of oxide stringers is not as dependent on the Y concentration in the Y-containing alloys. Furthermore, in the 0.1 % Y alloy, there are not sufficient present to completely eliminate multilayer scale formation as a contributory factor to the overall weight gain.

Under thermal cycling conditions, again two factors appear to be important. First, the presence of Hf or Y in the alloy and the resulting formation of oxide protrusions into the alloy at the scale/alloy interface lessens the likelihood of scale spallation. The number of stringers is important, and there are not sufficient in the alloys containing 0.05 and 0.1 % Hf: 0.3 and 1.0 % Hf additions appear to be optimal. Furthermore, Hf seems more efficient than Y and this is related to the shape of the pegs as much as the concentration. With the Hf additions, the internal growth of the Al_2O_3 around the Hf-rich internal oxide particles takes on a branched, dendritic form (Fig. 10) as opposed to the relatively smooth interface between the Al_2O_3 surrounding the Y-rich particles and the alloy. These differences may well be related to the distribution of the active element in the original alloy. Y tends to segregate to grain boundaries as an intermetallic yttride: Hf is completely in solid solution and this leads to a very fine distribution of small internal oxide precipitates which then promote the branching growth of Al_2O_3 around them. There is also some evidence to suggest that enrichment of Y may occur in the internal oxide zone promoting growth of larger particles: this does not seem to be the case with Hf.

Thus, small pegs seem to be the best, preferably fairly densely distributed; large pegs, in fact, appear to be capable of initiating scale failure themselves.

The second possible factor contributing to the improved resistance to thermal cycling promoted by the active element addition, is related to the lower overall growth rate of the Al_2O_3. This means that there is less depletion of aluminium from the alloy surface, and thus a greater chance of Al_2O_3 re-forming if scale loss occurs.

In conclusion, then, Hf seems preferable as an active element addition. The distribution of pegs is far more uniform and few large ones form. Alloys containing 0.3 or 1.0 wt % Hf were the best of all the alloys examined for cyclic oxidation resistance; under less severe conditions, such as isothermal exposure, the 0.05 and 0.1 wt % Hf alloys are better, since the formation of fewer pegs means that there is less localized thickening of the Al_2O_3 scale.

ACKNOWLEDGEMENTS

This work was initiated under the sponsorship of the Cobalt Information Centre and completed with the support of the University of Liverpool.

REFERENCES

1. C. S. Wukusick and J. F. Collins (1964). *Mat. Res. Stand.*, **4**, 637.
2. E. J. Felten (1961). *J. Electrochem. Soc.*, **108**, 490.
3. B. Lustman (1950). *Trans. TMS–AIME*, **188**, 995.
4. J. K. Tien and F. S. Pettit (1972). *Met. Trans.*, **3**, 1587.
5. J. Stringer (1966). *Met. Rev.*, **11**, 113.
6. H. Pfeiffer (1957). *Werkstoffe Korr.*, **8**, 574.
7. J. M. Francis and J. A. Jutson (1968). *Corros. Sci.*, **8**, 574.

8. G. R. Wallwork and A. Z. Hed (1971). *Oxid. Met.*, **3**, 213; also G. N. Irving, D. P. Whittle and J. Stringer, *Corrosion* (in press).
9. C. W. Price, I. G. Wright and G. R. Wallwork (1973). *Met. Trans.*, **4**, 2423.
10. C. S. Giggins and F. S. Pettit (1976). Final Rept to Aerospace Res. Labs, Wright Patterson AFB, Contract No. F33615-72-C-1702.
11. Y. Oishi and W. D. Kingery (1962). *J. Chem. Phys.*, **37**, 480.

DISCUSSION

J. A. Klostermann (University of Technology, Eindhoven, Netherlands): Do you think that the smooth regions are formed by vacancy condensation? And if this is the case, there must be a large influence of the Y or Hf additions on the point defect concentrations? Can you give an interpretation in terms of diffusion of point defects?

Reply: It is probable that the smooth regions are formed by vacancy condensation. However, the role of Y or Hf additions is not to alter the point defect concentrations but to provide alternative sites, away from the alloy/scale interface, for vacancy coalescence.

R. C. Hurst (Euratom, Petten, Netherlands): The apparently large growth stresses implied in the SEM of the oxidized alloy containing no Y or Hf implies that the stringers would have to be subjected to enormous stresses, much greater presumably than their fracture stress. Do the authors imply that the presence of Y or Hf reduces these growth or thermally induced stresses?

Reply: The convoluted scales on the alloys containing no Y or Hf are not a result of growth stresses. The scale when detached from the alloy grows in this manner due to continued transport of Al from the alloy via a vapour phase.

A. R. Nicoll (Brown Boveri, Heidelberg, Germany): Have the Y and Hf variations been compared in any form of turbine test?

Reply: As far as I am aware there have been no investigations directed at the comparison of Hf and Y additions in a more practically orientated turbine test. However, the main advantage of Hf additions over Y is in improving scale/metal adherence under oxidation conditions at relatively high temperature, and as such it is not anticipated that improvement in behaviour at lower temperatures in the somewhat more contaminated environment of an industrial turbine would be achieved.

5

Effect of Grain Size and Y_2O_3 Additions on the Hot Corrosion Behaviour of IN-738

P. HUBER

Sulzer Brothers Ltd, Winterthur, Switzerland

and

G. H. GESSINGER

Brown Boveri Research Centre, Baden, Switzerland

ABSTRACT

The influence of grain size and Y_2O_3 dispersion on hot corrosion behaviour was determined on three modifications of a nickel-base alloy IN-738:

(1) *Dispersoid-free IN-738 (cast, coarse-grained).*
(2) *Y_2O_3-dispersed IN-738 (mechanically alloyed, fine-grained).*
(3) *Y_2O_3-dispersed IN-738 (mechanically alloyed, intermediate grain size).*

The corrosion tests were carried out in a burner rig simulating the operating conditions of a stationary gas turbine at temperatures of 850 and 950°C. The corrosion rates and the composition of the corroded surface layers were determined and compared. The results show that the alloys containing a dispersion of Y_2O_3 have a lower corrosion rate at all temperatures considered. At 950°C a finer grain size further reduces the corrosion rate, whereas the effect of the dispersion on the corrosion rate is predominant at 850°C.

At 850°C the sulphide formation is reduced by the likely formation of yttrium oxysulphide; at 950°C where internal oxidation predominates the main effect of a small grain size is thought to be an increase in the rates of chromium and aluminium diffusion.

71

INTRODUCTION

The oxides of the rare-earth elements have long been considered to be beneficial for the oxidation resistance of nickel- and cobalt-base alloys (1–5). Oxide dispersions not only improve the corrosion resistance (6–9) in general but also increase the elevated temperature creep strength of such alloys. The mechanical alloying process as described by Benjamin (10) has greatly revived the interest in oxide-dispersion-strengthened alloys and in their high temperature properties.

The large number of papers dealing with rare-earth additions to high temperature alloys can be grouped into two types:

(1) Additions of rare-earth oxides and their effect on the oxidation and sulphidation behaviour.
(2) Additions of rare-earth metal only.

While the beneficial effect of rare-earth oxides on the corrosion behaviour is uncontested, the role of rare-earth metal additions is still questionable. In cases where a reduction of the corrosion rate has been reported, very often a prior oxidation step has led to a microstructure similar to those in oxide-containing alloys.

The purpose of the present investigation was to examine the influence of Y_2O_3 and alloy grain size on the hot corrosion behaviour of a nickel-base alloy IN-738. Three alloy modifications were compared:

(1) Dispersoid-free IN-738 (cast, coarse-grained);
(2) Y_2O_3-dispersed IN-738 (mechanically alloyed, fine-grained);
(3) Y_2O_3-dispersed IN-738 (mechanically alloyed, intermediate grain size).

The oxide-containing alloys were made as part of an earlier programme, where the major objective was to improve the high temperature creep rupture strength (11).

EXPERIMENTAL PROCEDURE

Alloys Used

The alloy preparation by mechanical alloying of IN-738 and Y_2O_3 has been described in great detail before (11). The alloys were made by blending high purity master alloys containing the reactive elements aluminium, titanium and zirconium (Ni/15Ti/15Al and Ni/29Zr) with elemental powders of Co,

Cr, Mo, W, Ta, Nb and Ni, as well as Y$_2$O$_3$ powder in an attritor. The milling operation was carried out under an argon atmosphere for a time of ca. 48 h. Mechanically alloyed powder was canned in mild steel containers and extruded at 1060 °C, using an extrusion ratio of 9:1.

The extruded bars were given two types of heat treatment. The fine-grained alloy was obtained by heating in argon at 1100 °C for 3 h, with a resulting grain size of 1–5 μm. The intermediate grain size alloy was made by heating the extruded bar at 1270 °C for 3 h. This annealing treatment resulted in an average transverse grain diameter of 40 μm and an average longitudinal grain diameter of 100 μm (the grain dimensions were obtained by lineal analysis). Both alloys were cooled in air and tested in this condition, i.e. no aging treatment was applied.

The cast alloy was supplied by International Nickel and had an average grain diameter of 3000 μm. The microstructure of the three alloys used can be seen in Fig. 1.

The main difference in the chemical composition of the alloys is the absence of boron, the reduced amount of carbon (0.077 as compared with 0.17 C for the cast alloy) and the presence of 1.5 wt % Y$_2$O$_3$ in the oxide-dispersion-strengthened alloys.

Corrosion Tests

The corrosion tests were carried out on a burner rig described elsewhere (12). With it the operating conditions of a stationary gas turbine, with the exception of pressure, can be simulated as closely as possible. Earlier work with this rig on a large number of high temperature alloys has demonstrated that the morphology of the resulting corrosion layers corresponds exactly with those observed in actual service. The most important operating data are summarized in Table 1. In order to accelerate the corrosive attack 15 ppm sodium and 5 ppm vanadium were added.

TABLE 1
Operating data of corrosion rig and fuel composition

Temperature	850 °C and 950 °C
Pressure	1.1 bar
Duration	at 850 °C—total 900 h (3 × 300 h)
	at 950 °C—total 300 h
Fuel	extra-light fuel oil
	(sulphur content 0.3–0.4 %)
Corrosive additions	15 ppm Na⎫ as naphthenates
	5 ppm V ⎭
Excess air	200 %

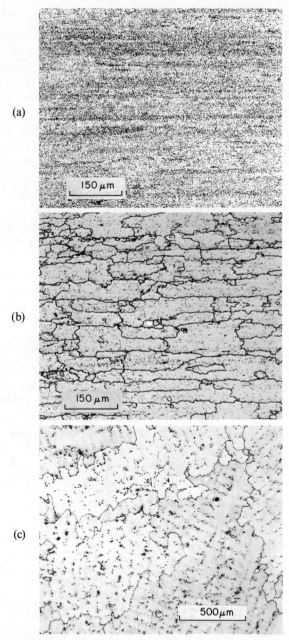

FIG. 1. Microstructure of alloy conditions used: (a) IN-738 + 1.5% Y_2O_3, annealed for 3 h at 1100 °C; (b) IN-738 + 1.5% Y_2O_3, annealed for 3 h at 1270 °C; (c) IN-738, cast

The specimens were provided in the form of cylindrical rods (8 mm diameter and 80 mm length) and were placed in an upright position in the ceramic lining of the burner rig.

As is now accepted practice in dynamic corrosion tests the corrosion rate was defined as the difference between the initial specimen thickness, measured prior to the test at various defined places, and the thickness of the material not yet attacked. This latter thickness measurement was always based on the maximum penetration depth at any point along the circumference of a sample.

In order to follow the effect of time on the corrosion progress at 850 °C, three test intervals of 300 h each were used and material for metallographic inspection was cut from the test bars.

The corroded test pieces were sectioned transversely and examined by light microscopy and by electron microprobe analysis.

RESULTS

Corrosion Tests at 850 °C

Figure 2 shows the penetration thickness as a function of time. It can be seen that both the fine-grained and the coarse-grained variants of the Y$_2$O$_3$-dispersed samples show a much better corrosion behaviour than the Y$_2$O$_3$-free cast alloy, which has been attacked about three times as much as the fine-grained oxide dispersion alloy after 900 h. All specimens show an increase in corrosive attack with increasing time which is mainly due to sulphidation, as will be seen later.

Figure 3 shows the surface regions of alloy IN-738 after 900 h at 850 °C. Comparing the cast alloy with the oxide dispersion alloys two observations can be made:

(1)   The interface between the surface scale and the uncorroded base material is much straighter and less interrupted in the oxide dispersion alloys.

(2)   The sulphide layer between the surface and the interior of the alloy is very distinct in the cast alloy, but it cannot be seen in either the fine-grained or the coarse-grained oxide dispersion alloys.

Figure 4 displays the electron microprobe analyser scans for the cast alloy (high corrosion rate) and the fine-grained oxide dispersion alloy (low

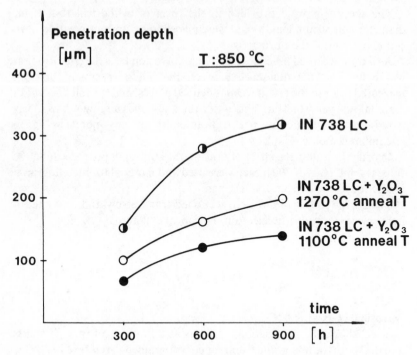

FIG. 2. Penetration thickness after corrosion at 850 °C as a function of time

corrosion rate). The elements displayed are chromium, aluminium and sulphur. Noteworthy in the cast alloy is the sulphidation zone between the scale and the undamaged interior and a layer of Al_2O_3 above the sulphidation zone. The surface scale consists mainly of Cr_2O_3, the spinel $NiCr_2O_4$ and rutile TiO_2. A completely different appearance is given by the oxide dispersion alloys. There is essentially no internal sulphur present, and the protective oxide layers are more uniform and are coherently distributed over the entire specimen. A wet chemical analysis of the surface scale gave an indication of small amounts of sulphur, which could not be detected in the surface scale or the cast alloy.

Figures 5(a) and 5(b) show the X-ray line scan of Cr, Ti and Al for the two oxide dispersion alloys. There is a layer depleted of γ' which is directly underneath the scale. The influence of grain size can be mainly seen in the Cr distribution. In the coarse-grained alloy there is a partial depletion of Cr underneath the scale but the bulk alloy Cr level is maintained in the fine-grained alloy.

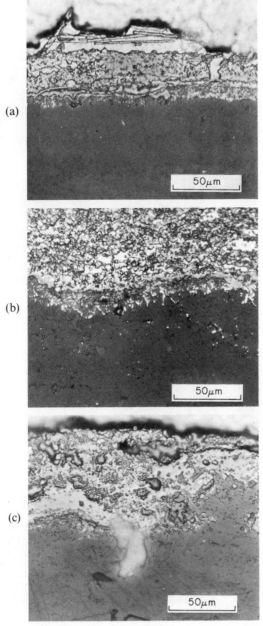

FIG. 3. Microstructure in the surface region of IN-738 after 900 h at 850 °C: (a) IN-738 + 1.5% Y_2O_3, annealed for 3 h at 1100 °C; (b) IN-738 + 1.5% Y_2O_3, annealed for 3 h at 1270 °C; (c) IN-738, cast

P. Huber and G. H. Gessinger

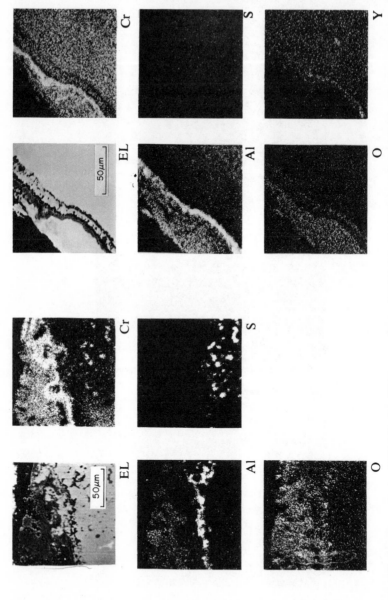

Fig. 4. Microprobe analysis of IN-738 after 900 h corrosion at 850 °C: left: IN-738, cast; right: IN-738 + 1.5% Y_2O_3, annealed for 3 h at 1100 °C

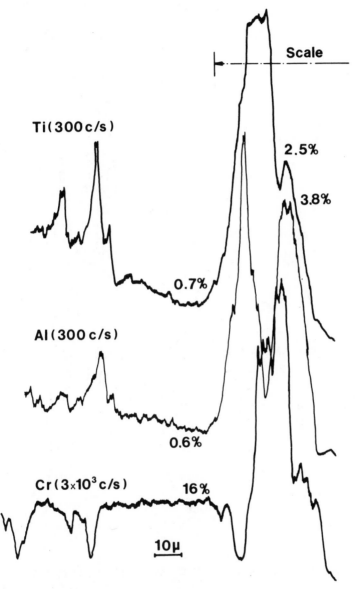

FIG. 5(a). X-ray line scan of Cr, Ti and Al: IN-738 + 1.5% Y_2O_3, annealed for 3 h at 1100 °C

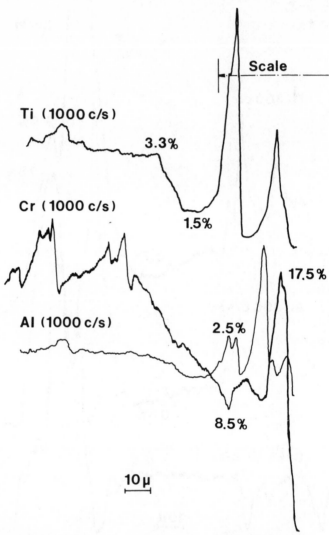

FIG. 5(b).   IN-738 + 1.5% Y_2O_3, annealed for 3 h at 1270°C

Corrosion Tests at 950 °C

Figure 6 shows the corrosion rate for 300 h at 950 °C. As at 850 °C the fine-grained oxide dispersion alloy shows the lowest corrosion rate, and the cast alloy the highest one.

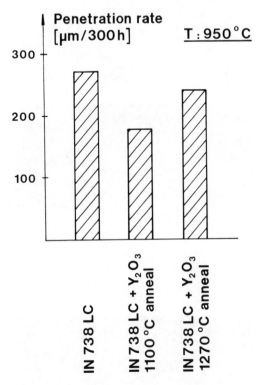

FIG. 6.   Corrosion rate over 300 h at 950 °C

Figure 7 shows the surface regions of alloy IN-738 after 300 h at 950 °C. Here the cast alloy has been compared with the two oxide dispersion alloys. There is no sulphur penetration visible (and no sulphur in the scale, as shown by wet chemical analysis). The Al_2O_3 layer is coherent in the fine-grained oxide dispersion alloy, whereas a discontinuous formation of Al_2O_3 in the cast alloy is observed. The coarse-grained oxide dispersion alloy is more similar to the cast alloy both in appearance of the Al_2O_3 reaction layer and in corrosion rate.

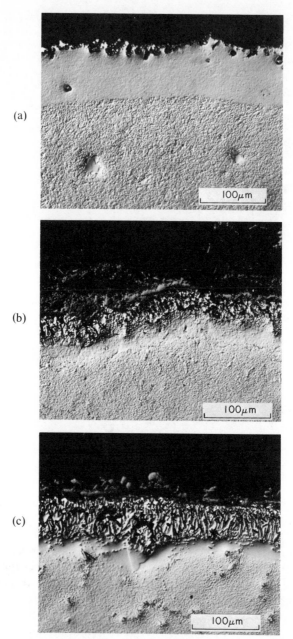

FIG. 7. Microstructure in the surface region of IN-738 after 300 h at 950 °C: (a) IN-738 + 1.5 % Y_2O_3, annealed for 3 h at 1100 °C; (b) IN-738 + 1.5 % Y_2O_3, annealed for 3 h at 1270 °C; (c) IN-738, cast

DISCUSSION AND CONCLUSIONS

The results of this investigation should help to indicate the effect of both grain size and Y_2O_3 dispersion on the hot gas corrosion rate at two temperatures for an engineering alloy under realistic testing conditions. In his classical work, Seybolt (6) has discussed the possible reactions of rare-earth oxide additions with sulphur. The most likely one to occur, which also should help to explain our results at 850 °C, is

$$Y_2O_3 + S_{in\ alloy} + 2/3\,Cr \rightleftarrows Y_2O_2S + 1/3\,Cr_2O_3$$

It is not certain that chromium is the only element which participates in the above reaction. Other strong oxide-forming elements, such as titanium or aluminium, would fill this role as well. The purpose of these elements is to provide additional driving force for the formation of the oxysulphide.

Since by this reaction most of the sulphur is gettered by the Y_2O_3 in the form of Y_2O_2S, no sulphur is available to react with either chromium or nickel to CrS or NiS, whose formation controls the rate of corrosive attack in the cast (dispersoid-free) alloy. Grain size of the alloy is of secondary importance. A reduced grain size does enhance the chromium diffusion to the outside, which improves the coherency of the oxide layer. The fact that corrosion at 850 °C occurs mainly by sulphidation and not by oxidation explains the large beneficial effect associated with the gettering action of the Y_2O_3.

The corrosion attack at 950 °C in our experimental conditions has changed in mechanism from predominantly sulphidation to predominantly internal oxidation. This is caused mainly by the increased rate of evaporation of Cr/Cr oxide. Seltzer and Wilcox (13) have shown that the rates of chromium and aluminium diffusion are increased by reducing the grain size of a nickel-base alloy but that they are not affected by the presence of an oxide dispersion directly.

An increase in the rate of aluminium diffusivity leads to the formation of a continuous Al_2O_3 layer in the oxide dispersion alloys, whereas the lower aluminium diffusivity in the coarse-grained cast alloy leads to the formation of a continuous Cr_2O_3 layer, together with an internally oxidized discontinuous Al_2O_3 layer. Since the oxygen diffusivity is greatly reduced by Al_2O_3, the alloy with a continuous Al_2O_3 layer shows the lowest corrosion attack.

REFERENCES

1. C. S. Giggins and F. S. Pettit (1971). *Met. Trans.*, **2**, 1071.
2. G. R. Wallwork and A. Z. Hed (1971). *Oxid. Met.*, **3**, 229.

3. H. H. Davis, H. C. Graham and I. A. Kvernes (1971). *Oxid. Met.*, **3**, 431.
4. I. G. Wright, B. A. Wilcox and R. I. Jaffee (1975). *Oxid. Met.*, **9**, 275.
5. H. T. Michels (1976). *Met. Trans.*, **7A**, 379.
6. A.U. Seybolt (1969). Final Report to Office of Naval Research Contract No. N00014-68-C-0362, Washington, DC 20360 (December).
7. I. G. Wright, B. A. Wilcox and R. I. Jaffee (1973). Final Report to NASA on Contract No. N00019-72-C-0190 (January).
8. W. Krajewski and H. Winterhager (1976). *Metall*, **30**, 441.
9. A. Raman (1977). *Z. Metallk.*, **68**, 163.
10. J. S. Benjamin (1970). *Met. Trans.*, **1**, 2943.
11. G. H. Gessinger (1976). *Met. Trans.*, **7A**, 1203.
12. P. Felix (1973). In *Deposition and Corrosion in Gas Turbines* (Ed. A. B. Hart and A. J. B. Cutler), CEGB Conf., Applied Science, London.
13. M. S. Seltzer and B. A. Wilcox (1972), *Met. Trans.*, **3**, 2357.

DISCUSSION

A. Naoumidis (Kernforschungsanlage Jülich, Germany): The microprobe analysis of the Y-containing alloys reveals that the Y is enriched at the metal/scale interface. Does this local high Y concentration form a diffusion barrier? Can this be an explanation for the low oxidation rate of the Y-containing alloys?

Reply: I think that the formation of Y_2O_3 between the oxide layer and the uncorroded metal is one of the reasons for the better corrosion behaviour. In the literature a double oxide ($YAlO_3$ and $Y_3Al_5O_{12}$) is known, which promotes oxide adherence and it may also act as a diffusion barrier to give better corrosion resistance.

R. Pichoir (ONERA 92320, Chatillon, France): You are comparing the oxidation behaviour of a Y_2O_3-dispersed alloy manufactured by powder metallurgy technique with that of a cast alloy devoid of Y_2O_3. I think it is difficult to draw conclusions about the role played by Y_2O_3 since the two materials are different in some other respects, i.e. segregation, grain size, etc. Don't you think that it is necessary to investigate the oxidation behaviour of the alloy obtained by powder metallurgy techniques but without the presence of Y_2O_3? If so, do you plan to undertake this study?

Reply: I think the corrosion behaviour of the alloy produced by PM techniques is different from that of the cast alloy. We used the corrosion rate of cast alloy as standard and we saw that at 850 °C for sulphidation attack the addition of Y$_2$O$_3$ gave a bigger benefit than the decrease in grain size, because the difference in the corrosion rates of the two PM alloys was small. At 950 °C, only the fine-grained Y$_2$O$_3$ dispersion alloy gives a significantly better corrosion behaviour; the difference in behaviour between the cast alloy and the large-grained PM alloy is very small in these conditions.

J. C. Colson (Laboratoire de Recherches sur la Réactivité des Solides, Dijon, France): Have you some data on the corrosion rate of the sintered IN-738 alloy? I have some results on this subject and it appears that the sintered alloy has a better behaviour than the cast alloy. The addition of dispersed particles may have a good effect but the preparation method of a sintered alloy may also change its corrosion behaviour. I would like to know your opinion.

Reply: I have no data on the corrosion rate of sintered IN-738 but I think too that the preparation by sintering changes the corrosion behaviour. Our investigation examined the effect of Y$_2$O$_3$ in combination with the effect of grain size. At higher temperatures (oxidation attack) the difference between the corrosion rates of the two Y$_2$O$_3$ dispersion alloys is bigger than at 850 °C where sulphidation attack is occurring. So it seems that at 950 °C the influence of the grain size is of more importance than the addition of Y$_2$O$_3$.

M. Lambertin (Laboratoire de Recherches sur la Réactivité des Solides, Dijon, France): You showed in the case of the alloys containing dispersed particles that the corrosion rate was dependent on the annealing temperature and consequently on the grain size. We observed the same phenomenon when studying the sulphidation of Fe–Cr alloys prepared by extrusion from powders with different sizes. Can you explain this phenomenon?

Reply: Your results confirm our theory for the oxidation attack. I cannot explain this phenomenon any better than you can. I think it must be some effect of short-circuit diffusion producing uniform protective chromium oxide layers; the chromium diffusion to the surface will be made easier by the smaller grain size.

6

Effect of High Temperature Corrosion on the Creep Behaviour of Aero Gas Turbine Materials Under Service-Like Conditions

Kh. G. Schmitt-Thomas, H. Meisel and H.-J. Dorn

Technische Universität, München, West Germany

ABSTRACT

Creep tests on the nickel-base materials René-41 and IN-100 under static stress in a stream of JP4 hot gas and in air show that there is a marked drop in creep strength in a corrosive medium. If NaCl is added to the fuel, a further drop in the creep strength is observed.

Metallographic, scanning electron microscope and microanalytical examination show similar alloy microstructures and protective scale formation under stress in air and in JP4 hot-gas atmosphere. However, the creep behaviour shows a specific dependence on the surrounding medium. The constant creep rate region of the creep curves is eliminated in a corrosive atmosphere. Microfractographic assessment of the fracture faces enables typical fracture features to be associated with the differing creep behaviour. In a corrosive medium, regions of a ductile phase form on the grain boundary faces and cause earlier fracture of the material.

To shed light on these specific corrosion mechanisms, test specimens were predamaged in hot gas and broken at service temperature under vacuum in an Auger microprobe. The results of these investigations have provided an interpretation of the hot corrosion mechanism in aero gas turbines.

INTRODUCTION

Aero gas turbine materials are exposed not only to purely thermal stresses but to a multitude of complex influences as well. These include slowly and rapidly fluctuating mechanical stresses, fluctuating thermal stresses and corrosive effects—especially in the form of gas–metal reactions in the reactive medium of the hot gas—caused by contamination of the fuel and of the intake air. When gas turbine components are designed, the most important criterion is the creep strength, based usually on the results of creep tests conducted in still air. However, the figures obtained for strength in still air have only limited applicability to the conditions inside a turbine, as the effects of erosion and corrosion are not covered in this kind of test.

This chapter reports the results of an investigation which had the objective of clarifying the effect of hot-gas corrosion on the creep behaviour of aero gas turbine materials under service-like conditions.

EXPERIMENTAL

It was possible to use an existing test facility of Motoren-und Turbinen-Union (MTU) in Munich for this purpose (Fig. 1). For the test to simulate service conditions as closely as possible, JP4 aviation fuel is burned in the combustion chamber of a MTU small gas turbine and the hot gas generated in this way is routed through a nozzle at a velocity of ~ 220 m/s on to the test specimen material. The test specimen can be loaded both statically—to determine the creep strength—and dynamically with slow and rapid load cycles, to determine the fatigue strength. More severe corrosion conditions

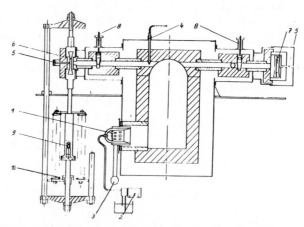

FIG. 1.   Hot-gas test bench (schematic): 1, combustion chamber; 2, fuel pump; 3, primary air; 4, chamber thermocouple; 5, specimen thermocouple; 6, creep specimen; 7, corrosion specimen; 8, temperature control; 9, strain gauge; 10, stress recorder

TABLE 1
Composition of the materials

Alloy	Composition in wt %								
	Al	Co	Cr	Mo	Ti	Zr	B	C	Ni
R-41	1.5	11	19	10	3.1	—	0.01	0.09	Bal.
IN-100	5.5	15	10	3	5	0.05	0.015	0.18	Bal.

are obtained by injecting artificial sea water through a cooled capillary tube
to produce a salt content of 1.3 ppm in the hot-gas stream. The materials
René-41 and IN-100 (Table 1) were examined.

RESULTS AND DISCUSSION

In comparison with the behaviour in hot air, the creep strength of the
materials dropped markedly in JP4 hot-gas atmosphere (Fig. 2), as shown
by the example of the alloy R-41. If artificial sea salt was also added to the
hot-gas atmosphere, a further drop in the creep strength resulted.

For comparison with the behaviour of the forging material R-41, the
drop in the creep strength of the precision-casting alloy IN-100 is shown in

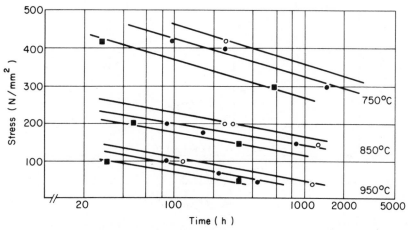

FIG. 2.   Creep behaviour of R-41 in air and JP4 combustion gases. ● JP4 gases; ○
air; ■ JP4 gases + salt

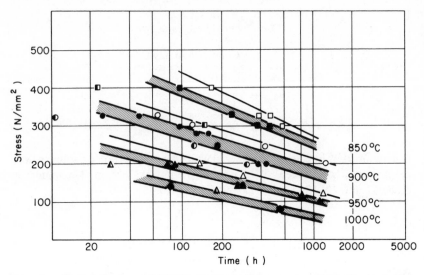

FIG. 3.   Creep behaviour of IN-100. ▣, ◑, ▲ JP4 gases + salt; ■, ●, ▲, ◆ JP4 gases; □, ○, △ air

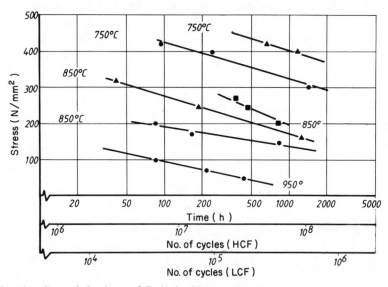

FIG. 4.   Creep behaviour of R-41 in JP4 combustion gases under static and dynamic stress. ▲ Low cycle (0.11 Hz) fatigue; ■ high cycle (20 Hz) fatigue; ● static stress

Fig. 3. The drop in strength is similar to that for R-41, but the creep rupture strengths are generally higher than those for the forging material R-41.

Testing of materials under dynamic stress shows that dynamic stresses cannot be considered significantly more critical for service life than static stresses. This statement can naturally only be made for the frequencies of 0.11 cps and 20 cps which were investigated. In Fig. 4 the effects of static stress on R-41 are compared with those of dynamic stress, all other thermal and corrosive conditions remaining the same. In the case of the low cycle fatigue (LCF) tests particularly, the life-reducing effect on the test specimen is mainly restricted to the periods when the load passes through its maximum phase. This distinction cannot be made quite so clearly in the case of the high cycle fatigue (HCF) tests.

After the corrosion tests under mechanical stress, the specimens were examined metallographically, paying special attention to the fracture morphologies and the microstructure. In addition, investigations of the protective scale were made. During the creep tests in air and JP4 hot gases, protective scales showing no dependence on the surrounding medium are

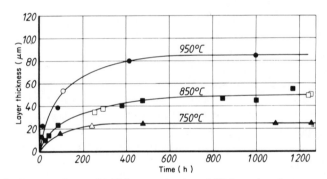

FIG. 5. R-41 protective scale thicknesses in air and JP4 combustion gases. ●, ■, ▲ JP4 gases; ○, □, △ air

formed. In Fig. 5 the growth curves for these protective scale layers on the alloy R-41 are shown.

Similar results were obtained for the thicknesses of the protective scale layers formed on IN-100 specimens during creep testing (Fig. 6).

A key to protective scale behaviour and the kinetics of protective scale formation is a knowledge of the composition of the protective scale layers. A wavelength dispersion electron microprobe analysis was made of the concentration profiles of the protective scale-forming elements, shown here using the alloy R-41 as an example.

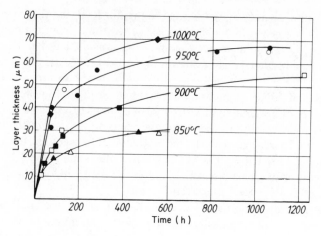

FIG. 6.   IN-100 protective scale thicknesses in air and JP4 combustion gases. ◆,
●, ■, ▲ JP4 gases; ◇, ○, □, △ air

A comparison has been made of the concentration profiles in the protective scale layers on creep test specimens which were tested in air (Fig. 7) and JP4 combustion gases (Fig. 8) under the same thermal and mechanical stresses. The protective scale build-ups in air and JP4 combustion gases are comparable.

In contrast to the protective scale layer growth illustrated in Figs. 7 and 8, spalling of the protective scale takes place in combustion products to which sea salt has been added; this leads to a marked reduction of the cross-section and thus to premature initiation of fracture mechanisms in the material (Fig. 9).

Examination of the microstructure shows that the kinetics of precipitation in the material are dependent on the alloy composition but are independent of the surrounding atmosphere. The changes found in the structure, e.g. modification of carbides and growth of the γ'-phase (Fig. 10), are determined by temperature and time of exposure. Often, an oriented growth of the γ'-phase dependent on the stress direction is observed in the zone near the fracture.

The creep behaviour of the materials, on the other hand, is clearly dependent on the surrounding medium. In air, there is a lengthy constant creep rate regime but this is severely reduced in JP4 combustion gases (Fig. 11).

The constant creep rate regime was almost completely eliminated in

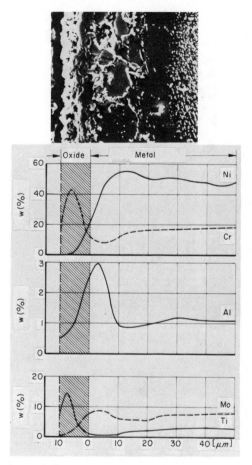

FIG. 7. Protective scale layer on a creep specimen tested in air: material, R-41; stress, 100 N/mm²; temperature, 950 °C; time to rupture, 111 h

combustion products to which sea salt had been added. This characteristic behaviour was shown by both materials (Fig. 12).

Valuable information on failure mechanisms is obtained from the microfractographic examinations. In combustion products, an increased tendency to the formation of microcracks can be observed. In air and JP4 combustion gases, the test specimens show mainly intercrystalline fracture. In air, the fractures are free of surface layers and smooth grain boundary faces are obtained (Fig. 13).

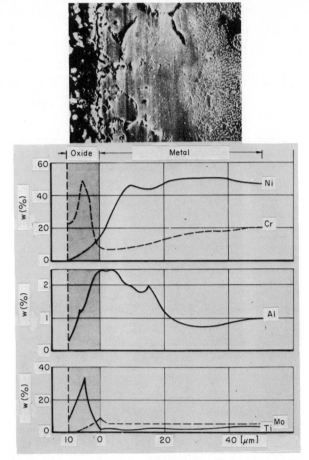

FIG. 8.   Protective scale layer on a creep specimen tested in JP4 combustion gases:
material, R-41, stress, 100 N/mm²; temperature, 950 °C; time to rupture, 87 h

In a corrosive atmosphere, on the other hand, fracture characteristics
specific to the medium can be observed on the grain boundary faces (Figs.
14 and 15). For example, on the intergranular fracture surfaces, films
showing ductile fracture can be observed.

The grain boundary surface coatings seem to be the result of reaction
between the material and the medium. The morphology indicates that these
grain boundary surface coatings have a low melting point and thus lead to
accelerated creep and premature fracture of the material (Fig. 16).

FIG. 9. Protective scale spalling on a creep test specimen: material, IN-100; temperature, 950°C; medium, JP4 + NaCl; time to rupture, 28 h

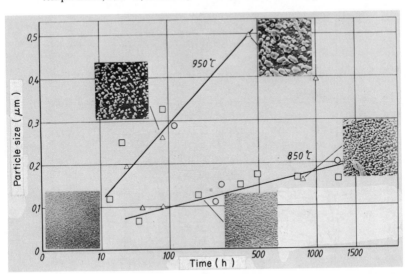

FIG. 10. Growth of the γ'-phase under creep stress. △ Static stress in combustion gases; ○ static stress in air; □ dynamic stress in combustion gases

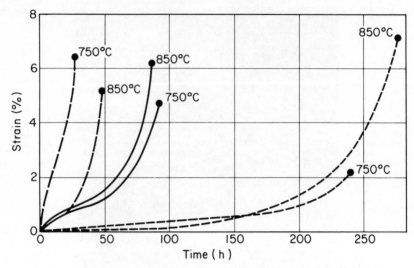

FIG. 11.   Creep behaviour of R-41 in JP4 gases + NaCl———; in JP4 gases——;
and in air -----. Stress 420 N/mm² at 750 °C; 200 N/mm² at 850 °C

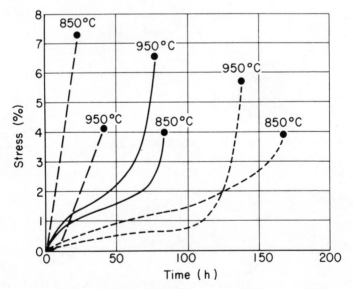

FIG. 12.   Creep behaviour of IN-100 in JP4 gases + NaCl———; in JP4 gases——;
and in air ----. Stress 400 N/mm² at 850 °C; 200 N/mm² at 950 °C

FIG. 13. Fracture face in a R-41 creep test specimen: material, R-41; temperature 750 °C; stress, 420 N/mm²; time to rupture, 242 h; medium, air

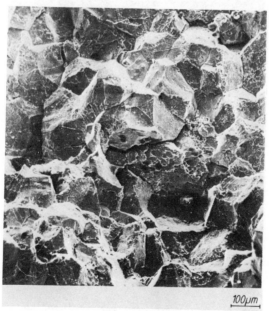

FIG. 14. Fracture face of a creep test specimen in JP4 hot-gas atmosphere: material, R-41; temperature, 750 °C; stress, 420 N/mm²; time to rupture, 94 h

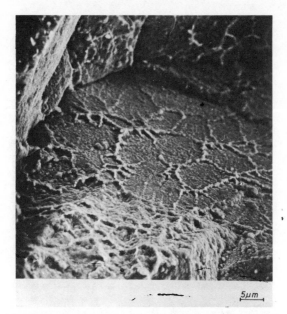

FIG. 15. Intergranular fracture with grain boundary face coating

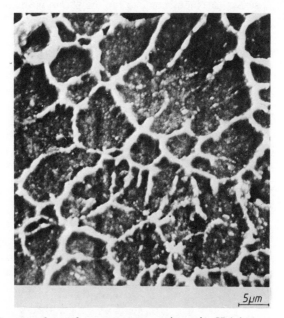

FIG. 16. Fracture face of a creep test specimen in JP4 hot-gas atmosphere: material, R-41; stress, 420 N/mm^2; temperature, 750°C; time to rupture, 94 h

The grain boundary fracture surfaces obtained under dynamic stress (Fig. 17) show the characteristic fatigue ripple markings.

Owing to the higher test temperature and the resulting increased scale formation on the fracture faces, the fracture features illustrated for R-41 could not be seen so clearly on alloy IN-100. However, in a test at 750 °C,

FIG. 17. Ripple markings after high cycle fatigue test in JP4 combustion gases: material, R-41; frequency, 20 cps; temperature, 850 °C; time to rupture, 809 h

comparable fracture behaviour in a corrosive atmosphere was found (Fig. 18).

Another objective of the investigation, therefore, was to determine the composition of the ductile phase on the grain boundary surfaces. Direct analysis of the grain boundary surface films by energy and wavelength dispersion electron probe microanalysis can be invalidated by the secondary reactions occurring as the stream of hot gas impinges on the fracture face after the fracture.

To avoid these difficulties, test specimens were therefore predamaged in the combustion products and then broken at service temperature in the high vacuum of an Auger microprobe.

The first microfractographic observations show a fracture appearance

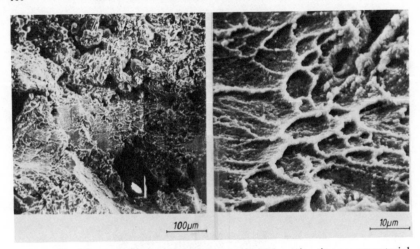

FIG. 18. Fracture face of a creep test specimen in JP4 combustion gases: material, IN-100; stress, 500 N/mm^2; temperature, 750 °C; time to rupture, 142 h

FIG. 19. Fracture surface of a predamaged test specimen broken under high vacuum: material, R-41; temperature, 750 °C

similar to that of the fracture faces produced in combustion products (Fig. 19).

In the region of the ductile grain boundary phase illustrated the Auger electron spectrum shows clear sulphur peaks (Fig. 20), chiefly near the surface.

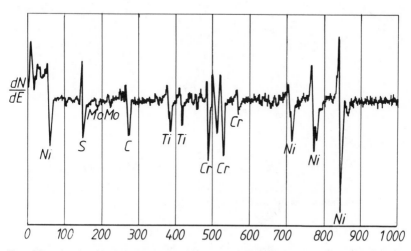

FIG. 20. Auger electron spectrum of a fracture face obtained under high vacuum after predamage in combustion products: material, R-41. Abscissa, electron energy (eV)

No sulphur enrichment was observed in the test specimens which were predamaged in air and broken under high vacuum (Fig. 21), so that the sulphur constituents in the combustion products must have been responsible for the sulphiding corrosion mechanism seen in the materials subjected to creep in this environment.

To sum up, it was shown that a distinct drop occurred in the creep strength of the materials R-41 and IN-100 in combustion products, as compared with their strengths in air. Furthermore, this drop in creep strength was even bigger if salt were added to the combustion products. It was shown that protective scales grew on the alloys exposed in air and JP4 combustion gases, but that this did not occur in atmospheres to which salt had been added.

Furthermore, the characteristic creep curves for the materials which were subjected to stress in differing media were examined. The fracture features found can be related to specific features of the curves. It is apparent that

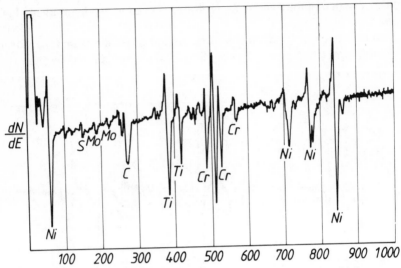

FIG. 21.   Auger electron spectrum of a fracture face obtained under high vacuum
after predamage in air: material, R-41. Abscissa, electron energy (eV)

sulphur constituents in the fuel lead to the formation of ductile phases on
the fracture faces and cause accelerated creep and thus earlier fracture of
the material.

DISCUSSION

H. F. Schmitz (KWU, Mülheim, Rühr, Germany): Under the COST-50
programme, Kraftwerk-Union has for some time been running creep tests
in air, in a corrosive atmosphere (air + 0.03 % SO_2/SO_3 + artificial ash), as
well as in a corrosive atmosphere to which chlorides have been added. A
marked difference in the behaviour of the materials has been observed.
Whereas Udimet 520 shows a big drop in creep strength in a corrosive
atmosphere, IN-738 turns out to be much less prone to high temperature
corrosion.

Have you, in your investigations, also found such marked differences
between the various materials, in the reduction of their creep strength by
corrosion?

Reply: Under various programmes, a lot of nickel-base materials have
already been tested in the MTU-München test facility I have described.

These include MAR-M246, MAR-M002, Nimonic 90, René-41, IN-100 and IN-713C. The lower creep strength in JP4 combustion gases than in air, first verified by Dr Huff, was observed in all materials. The drop in creep strength is comparable for the materials mentioned, when allowance is made for the usual scatter in creep tests.

W. Betz (MTU-München, Germany): Random tests carried out on the MTU combustion gas corrosion test facility show that JP4 combustion products have the same effect in reducing the creep strength of IN-738 flat test specimens as they have on other materials.

7

Comparative Studies of Corrosion Due to Hot Salts or Combustion Products of Kerosene Containing Impurities

Y. Bourhis and C. St. John
l'Ecole Nationale Supérieure des Mines de Paris, France

and

R. Morbioli and S. Ferre
SNECMA, France

ABSTRACT

An investigation of hot corrosion of various nickel-base superalloys has been carried out using both salt-coated and burner-rig techniques for comparison purposes. Testing results were found to correlate to the extent that the corrosion behaviour of the alloys followed the same trends in both tests. However, some differences were found with respect to the mechanisms by which the corrosion attack takes place in the two environments. The burner-rig test simultaneously develops several mechanisms (sulphidation/oxidation, fluxing) which, in contrast, were detected separately or sequentially in the salt-coated test. These differences are thought to be due to the dynamic aspects (thermal cycling, gas speed, continuous heavy contamination) of the burner-rig test as compared with the static salt-coated test.

INTRODUCTION

Turbine blade hot corrosion has been and still is the subject of much research in the USA as well as in Europe. But it seems that the great number of results increases the contradictions and the disagreements, perhaps because of the number of parameters controlling the corrosion process. In

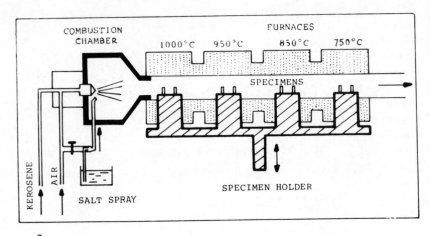

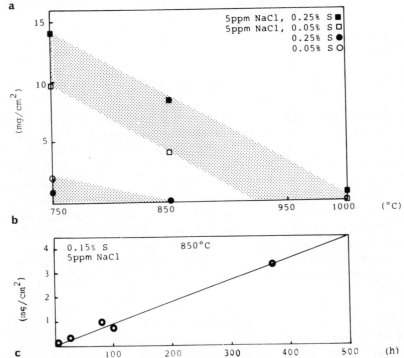

FIG. 1.   (a) Schematic diagram of the SNECMA burner rig; (b) amount of Na_2SO_4 deposited as a function of temperature after 500 h; (c) amount of Na_2SO_4 deposited as a function of time

fact, experimental approaches can be separated into two categories: mechanisms of corrosion studies which require a simple test with perfectly defined conditions; on the other hand, if the aim of the research is to compare alloy behaviour, then a realistic test will impose a number of experimental parameters more or less easy to control.

The purpose of this study is to show that (1) if the test conditions are well known, then thermobalance and burner-rig results do agree; and (2) thermobalance experiments isolate each mechanism of corrosion and, therefore, allow a more complete understanding of burner-rig experiments.

EXPERIMENTAL

The results and comparisons we obtained on a number of superalloys will be illustrated by those for two of them chosen for their typical behaviour: IN-738 and IN-100. As a result of their chromium and aluminium concentrations the former is essentially protected by a Cr_2O_3 scale and the latter by an Al_2O_3 scale.

Precontaminated samples are used in the thermobalance experiments: these samples are coated with Na_2SO_4 (from 0.05 to 5.0 mg/cm^2).

In the burner rig the sample environment depends on the conditions of the test. In Fig. 1(a) are given the description and characteristics of the burner rig used. The combustion chamber is fed with kerosene more or less contaminated with sulphur, and with air contaminated with a NaCl aerosol. Samples are placed in four temperature-controlled furnaces on a moving sample holder so that they can be thermally cycled.

On the whole, the environment of the samples is oxidizing and generally there is a deposit of sodium sulphate on them; very rarely it is possible to detect sodium chloride but, in these cases the ratio $NaCl/Na_2SO_4$ is less than 1/1000.

In Fig. 1(b, c) we can also see that the amount of Na_2SO_4 deposited increases roughly linearly with time and decreases with increasing temperature. The amount of deposit depends more closely on the concentration of sodium than on that of sulphur. Two types of test have been done: in the first test, with an addition of 0.25 % S to the kerosene, Na_2SO_4 was not detected on samples either at 1000 °C or at 850 °C; in the second, with 0.25 % S in the kerosene and 5 ppm NaCl in the combustion air, sulphate is present at both temperatures.

As regards the results themselves, it appears that, in the burner rig, corrosion is not always uniform and, for this reason, it seems interesting to note the corroded depths corresponding to the maximum and minimum values found after 5, 25, 125, 250 and 500 h testing.

RESULTS AND DISCUSSION

A Cr_2O_3-forming Alloy: IN-738

As mentioned previously (1), thermogravimetric curves obtained at 900 °C for 24 h show that with samples contaminated with 0.5 mg/cm² Na_2SO_4, there are generally two successive periods (see Fig. 2): a first interaction period between Na_2SO_4/oxide during which a weight loss is noted, probably due to the evaporation of sodium-containing double oxide and the liberation of SO_2/SO_3; and a second period during which the gain of

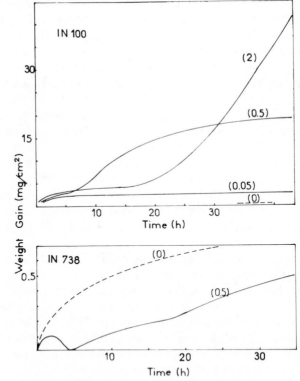

FIG. 2.   Corrosion kinetics at 900 °C with or without sodium sulphate coatings (mg/cm² of Na_2SO_4 is indicated in parentheses)

weight follows an approximatively parabolic kinetic law as with a non-contaminated sample. In this period the morphology of the oxide scales is also similar to that observed with a non-contaminated sample. However, the cubic structure of Cr_2O_3 co-exists sometimes with the usual hexagonal form and often, especially along the grain boundaries, there is slight precipitation of chromium and/or titanium sulphides between the internal oxidation zone and the matrix.

Higher contaminations and oxidation times up to 192 h do not essentially modify these conclusions; likewise, the behaviour remains very similar up to 1000 °C. This quasi-immunity towards sodium sulphate can be explained, according to Goebel *et al.* (2), by the buffering action of a Cr_2O_3/CrO_4^{2-} couple and by the physical barrier of Cr_2O_3.

The burner-rig results obtained with 0.25 % of sulphur are shown in Fig. 3(a). Since no Na_2SO_4 was detected on the samples, no corrosion appears either in the kinetic measurements or in the morphological examinations. The maxima and minima envelopes simply demonstrate the different metal losses at the grain boundaries and from the grains.

In Fig. 3(b) are shown results obtained with sodium sulphate deposits on samples when 5 ppm of sodium chloride was added to the combustion air. Compared with the preceding curves only a slight increase in rate appears at 850 °C; on the contrary, at 1000 °C the corrosion depth is more than twice as great.

The morphological examination (see Fig. 4) shows at 850 °C the appearance of an almost continuous layer of oxides composed essentially of $Ni(Cr, Al)_2O_4$, Cr_2O_3, Al_2O_3 and TiO_2, while internal sulphidation involving chromium, titanium and aluminium is visible in the perturbed underlying zone and in the grain boundaries. In Fig. 5 it can be seen that the oxide scale formed at 1000 °C lost its continuous nature, although it has the same elemental composition. The internal oxides and sulphides are very numerous. The corrosion propagates along the grain boundaries which are, in fact, diffusion short circuits and local reducing sites having the appropriate kinetic and thermodynamic conditions to promote sulphidation. Afterwards, the corrosion progresses inside the grains but the attack keeps its local nature.

At 1000 °C, there is no good agreement between the thermobalance and burner-rig results. This difference cannot be completely explained by the dynamic character of the burner-rig test which promotes the volatilization of Cr_2O_3, nor by the presence of SO_2/SO_3 in the environment because their dilution makes sulphidation thermodynamically unlikely. In compensation, the mixed action of chloride vapours (resulting from incomplete

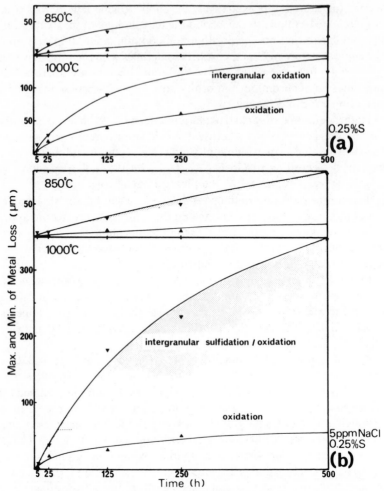

FIG. 3.   Burner-rig results obtained on IN-738 with (a) 0.25 % S and (b) 5 ppm NaCl, 0.25 % S in the gas stream

conversion into sulphate) and thermal shock can lead to the spalling of the oxides and, therefore, to the direct attack of the alloy by the liquid sodium sulphate. Hancock *et al.* (3, 4) demonstrated that each one of these factors is sufficient to provoke oxide spalling. Moreover, taking into account that the oxide adherence and the thermal cycling resistance both usually decrease with increasing temperature, then the difference at 1000 °C between the two tests can be explained.

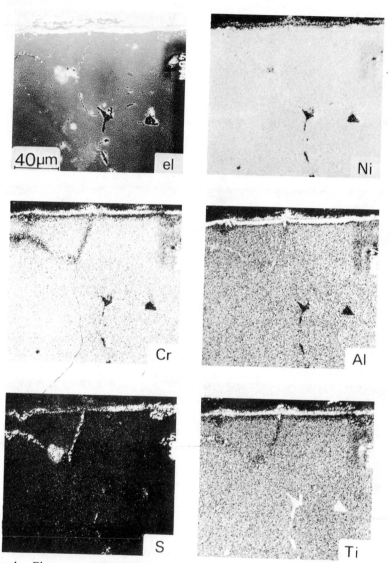

FIG. 4.   Electron and X-ray images of IN-738 oxidized at 850 °C for 250 h in the burner rig (5 ppm NaCl, 0.25 % S)

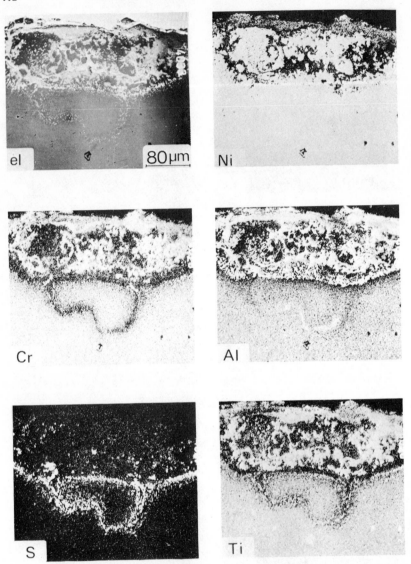

FIG. 5.   Electron and X-ray images of IN-738 oxidized at 1000 °C for 250 h in the
burner rig (5 ppm NaCl, 0.25 % S)

An Al_2O_3-forming Alloy Containing Molybdenum: IN-100

Previous results (1, 5, 6) showed that thermogravimetric curves obtained, after 24 h of heating at 900 °C, with 0.5 mg/cm^2 Na_2SO_4-contaminated samples, usually present three periods: 'incubation', quasi-parabolic and quasi-linear corrosion periods (see Fig. 2). To these three types of behaviour correspond three different and successive corrosion mechanisms.

First, there is an acidic fluxing of the Al_2O_3 layer by Na_2SO_4 according to the Goebel *et al.* (2) model, related to the equilibrium

$$SO_4^{2-} \rightleftharpoons SO_3 + O^{2-}$$

Then occurs a sulphidation/oxidation period leading to the building of a relatively compact oxide layer with, underneath, a molybdenum-rich zone. Finally, during the third stage, the formation of molybdenum-rich liquid oxides induces an acidic fluxing probably controlled by an equilibrium such as

$$MoO_4^{2-} \rightleftharpoons MoO_3 + O^{2-}$$

and a porous, powdery oxide is formed. This last corrosion mechanism roughly agrees with the interpretation proposed by Brenner (7), Peters *et al.* (8) and Fryburg *et al.* (9).

Moreover, it has been demonstrated (5) that the changes in test temperatures between 800 and 1000 °C do not drastically change these conclusions. On the contrary, the amount of contaminant strongly influences corrosion rates and mechanisms. For very low contamination (~ 0.05 mg/cm^2), only sulphidation/oxidation mechanisms appear. For contaminations greater than 5 mg/cm^2, sulphidation/oxidation replaces molybdate and sulphate fluxing as the general corrosion mechanism, but the latter can still produce severe local attack.

The burner-rig test results obtained with 0.25 % S are shown in Fig. 6(a). At 1000 °C no sulphidation is observed, as would be expected from the absence of Na_2SO_4 deposition on the samples; but at 850 °C the important difference between minimum and maximum penetration depths indicates the occurrence of non-uniform and unpredictable attack; for example, the 500 h sample was not corroded at the end of the test. These local attacks are associated with the precipitation of chromium-, titanium- and aluminium-rich oxides in a nickel matrix and a front of complex sulphides preceding these oxides. The presence of a superficial, almost continuous, Al_2O_3-rich layer indicates that the corrosion was developing under a very low partial pressure of oxygen and sulphur and for this reason only very reactive elements are involved. Corrosion began from very local failures due rather

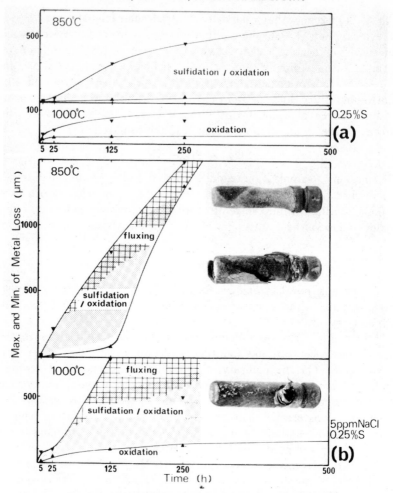

FIG. 6.   Burner-rig results obtained on IN-100 with (a) 0.25 % S and (b) 5 ppm
NaCl, 0.25 % S in the gas stream

to cracks than to fluxing of the protective layer. The observed sulphidation
shows that, although not detected, some sulphate must have been
deposited. In the thermobalance test, little sulphide was observed and this
only at temperatures below the sodium sulphate melting point; at these
temperatures the fluxing mechanism was very slow.

At 850 °C, when 5 ppm of NaCl was added to the combustion air, the
amount of sodium sulphate deposit is important and quickly reaches a level

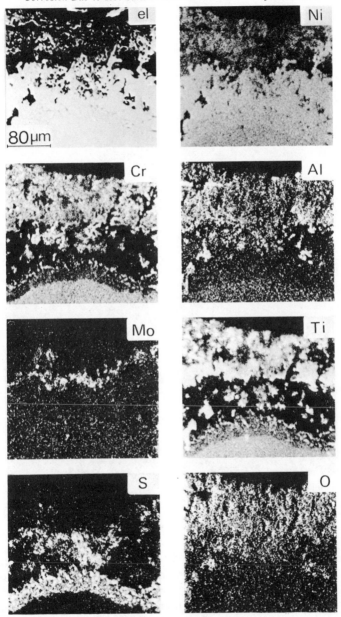

FIG. 7. Electron and X-ray images of IN-100 oxidized at 850 °C for 250 h in the burner rig (5 ppm NaCl, 0.25 % S)

116   *Y. Bourhis, C. St. John, R. Morbioloi and S. Ferre*

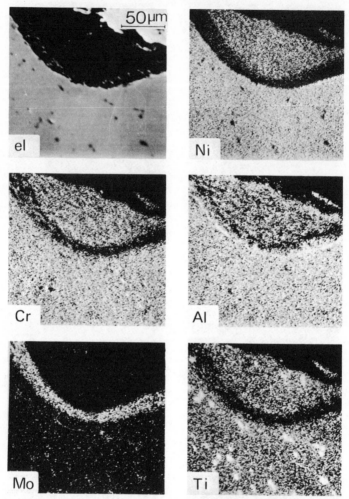

FIG. 8. Electron and X-ray images of IN-100 oxidized at 850 °C for 250 h in the
burner rig (5 ppm NaCl, 0.15 % S)

at which the alumina layer can be fluxed. The curves shown in Fig. 6(b)
show that, in these cases, corrosion progresses with successive periods of
sulphidation/oxidation and fluxing. This alternating attack systematically
induces the sample's destruction. On the contrary, at 1000 °C, the samples
can remain almost unattacked after 500 h or can be subjected to rapid
corrosion by one or both of these two mechanisms.

We present some photographs that illustrate different corrosion aspects corresponding to different types of destruction. If we keep in mind the temperature and time dependence of sulphate deposition (Fig. 1), then there is a rather good agreement with thermobalance results. There is also good agreement on micrographic aspects: the structure shown in Fig. 7 corresponds to a burner-rig corroded sample, but this morphology is very typical of that observed on a thermobalance-tested specimen during the parabolic kinetic period (5), namely molybdenum oxide precipitation preceded by a sulphide front. This morphology alternates during the test and along the sample surface with the structure shown in Fig. 8. The latter is characteristic of the phenomena observed in the thermobalance test during the linear kinetic period (5); the elements are uniformly distributed in a porous oxide which is isolated from the substrate by a continuous layer of molybdenum-rich oxide.

However, in spite of these similarities we must admit that fluxing corrosion mechanisms are less frequent in burner-rig than in thermobalance tests. Further, burner-rig results (5, 6) obtained on two other molybdenum-containing alloys (IN-713LC and MarM-246) showed sulphidation rather than a fluxing mechanism. Probably the reasons are: (1) the thermal cycling and high gas velocity which promote oxide spalling and volatilization of low melting point compounds, respectively, and (2) the effect of the heavy Na_2SO_4 deposit, which can produce high sulphur and low oxygen activities at the salt/oxide interface.

CONCLUSIONS

On two typical alloys it has been shown that corrosion induced by Na_2SO_4 can be reproduced by both burner-rig and thermobalance tests. Moreover, studies of micrographic structure demonstrate that the corrosion phenomena are very similar to those found in service on corroded blades cast from Cr_2O_3- and Al_2O_3-forming alloys. However, it must be mentioned that in these real cases, often sulphidation/oxidation mechanisms occur whereas signs of fluxing are rather rare. In fact, in practice, blades can be removed before the fluxing phenomenon appears.

The agreement between the results of the two types of test is rather satisfactory if we take into account that the burner-rig test is dynamic (thermal cycling, air turbulence and continuous Na_2SO_4 deposition). Moreover, samples can be oxidized before they are covered with a noticeable Na_2SO_4 deposit.

However, it must be emphasized that the burner-rig test gives less reproducible results than the thermobalance test, and also gives complex corrosion structures on the same sample due to the co-existence of different mechanisms (or different stages of one mechanism). For these reasons, the much simpler and easier to control thermobalance test is more suitable for mechanistic studies. As it gives quantitative information it therefore provides a good complement to the burner-rig tests.

REFERENCES

1. Y. Bourhis and C. St John (1975). *Oxid. Met.*, **9**, 507.
2. J. A. Goebel, F. S. Pettit and G. W. Goward (1973). *Met. Trans.*, **4**, 261.
3. P. Hancock (1974). Materials Science Symp., TMS–AIME Fall Mtg, Detroit, Michigan.
4. R. C. Hurst, J. B. Johnson, M. Davies and P. Hancock (1973). In *Deposition and Corrosion in Gas Turbines* (Eds A. B. Hart and A. J. B. Cutler), Applied Science, London, p. 143.
5. Y. Bourhis and C. St John (1976). In *Properties of High Temperature Alloys* (Eds Z. A. Foroulis and F. S. Pettit), The Electrochemical Society, New Jersey, p. 595.
6. Y. Bourhis and C. St John, unpublished.
7. S. S. Brenner (1955). *J. Electrochem. Soc.*, **102**, 16.
8. K. R. Peters, D. P. Whittle and J. Stringer (1976). *Corros. Sci.*, **16**, 791.
9. G. C. Fryburg, F. J. Kohl and C. A. Stearns (1976). In *Properties of High Temperature Alloys* (Eds Z. A. Foroulis and F. S. Pettit), The Electrochemical Society, New Jersey, p. 585.

DISCUSSION

R. C. Hurst (Euratom, Petten, The Netherlands): I would like to stress the importance of specifying the environment in rig testing, as it is of critical concern whether NaCl injected from sea water is converted to Na_2SO_4. This is particularly true when alloys with different susceptibilities to these two salts such as IN-100 and IN-738 are tested. For example, an overestimate of the corrosion resistance would be obtained if either IN-738 were to be subjected to Na_2SO_4 alone or IN-100 subjected to NaCl alone. However, it is clear that the same test might not produce the normal engine behaviour for a whole range of alloys and I would suggest that the laboratory test may have to be suited and adjusted only to one alloy at a time for satisfactory comparisons with engine experience.

Reply: The authors' experience agrees with the remark of Dr Hurst. We, too, have observed inverse behaviour when an Na_2SO_4-rich environment is replaced by an NaCl-rich one, with salt-coated tests as well as with burner-rig tests. Concerning the test reported in the present study, the burner rig operates with sodium chloride and sulphur as sources of contaminants, but almost total conversion into sulphate occurs, and this appears to be the most realistic condition, confirmed by analogous behaviour found on engine blades.

Concerning Dr Hurst's suggestion, it is certainly true that simulation tests would be better still if we could modify the environment, particularly the 'critical conditions' which differ from alloy to alloy. But if we adopted these experimental conditions we should lose, at least partly, the benefit of the comparison between alloys or we should have to make too many tests.

8

Corrosion Above 700°C in Oil-Fired Combustion Gases

S. Brooks, J. M. Ferguson, D. B. Meadowcroft and C. G. Stevens

Central Electricity Research Laboratories, Leatherhead, UK

ABSTRACT

The corrosion resistance of various materials has been investigated between 720 and 1250°C in a residual oil-fired environment using a burner rig. The materials included silicon-based ceramics and Fe–Cr–Ni, Fe–Cr and Cr–Ni alloys. The silicon-based ceramics were severely corroded when sodium sulphate deposition occurred (820–1100°C), as were the metal alloys which contained nickel; with the latter materials the attack decreased as the chromium content of the alloys increased. Except for 50Cr/50Ni alloys, the metals also suffered from a vanadium/sodium induced accelerated attack at lower temperatures (720–820°C). The mechanisms of the corrosion processes identified are discussed.

INTRODUCTION

In modern power station boilers there is a need for materials which can withstand the extremely arduous environment at temperatures in excess of 700°C. They are required for such items as burner components, tube spacers and hangers. This chapter describes an investigation of some corrosion-resistant materials, which are in the main commercially available and suitable for such applications in residual oil-fired boilers. The materials tested were Fe–Cr, Fe–Cr–Ni and 50Cr/50Ni-based alloys, and high strength silicon-based ceramics. In addition to reporting the loss of cross-section due to the corrosion of each material, the chapter emphasizes the corrosion mechanisms concerned, and the ways in which an understanding of these suggests more resistant material compositions.

EXPERIMENTAL

Most of the experimental work was carried out in an oil burner rig which had previously been found to simulate satisfactorily oil-fired boiler conditions. Figure 1 shows a schematic plan view of the rig. The combustion products from the oil burner passed at $\sim 5\,\mathrm{ms}^{-1}$ along the

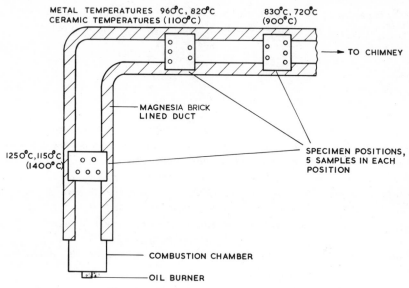

FIG. 1.   Plan view of the burner rig

15 cm square cross-section duct lined with magnesia bricks. Up to five samples could be hung across the duct in each of three positions, giving three exposure temperatures in each run. By varying the oil flow to the burner these temperatures could be adjusted. Residual oil-firing conditions were reproduced by adding sodium and vanadium naphthenates to distillate oil (to give 50 ppm V and 50 ppm Na in the fuel) and sulphur dioxide to the secondary air (simulating an oil with $\sim 3\%$ S). The rig operated with 2–4% excess oxygen.

The metal samples were rods 1.2 cm in diameter, 15 cm long, welded to support tubes as shown in Fig. 2. The ceramic samples were much smaller, usually in the form of bars 80 mm × 5 mm × 5 mm, and were suspended in the duct by platinum wire (Fig. 3). Hence, while the ceramic samples

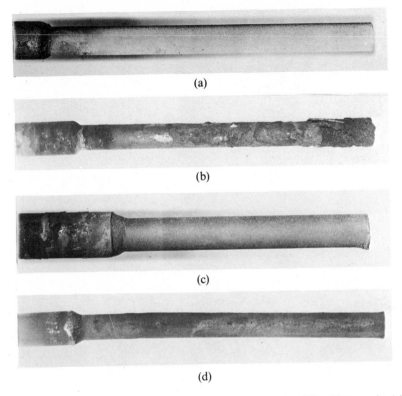

FIG. 2.   Test rods (12 mm diam.) after 225 h exposure in the oil-fired burner rig: (a) Kanthal (mod.) 1250 °C, high temperature protective scale; (b) AISI 310 960 °C, severe hot corrosion blistering; (c) Inconel 671 960 °C, hot corrosion at the end of a rod; (d) Inconel 589 1250 °C, deformation of rod in the gas stream

reached the gas temperature, thermal conduction along the metal rods kept these between 150 and 200 °C lower than the gas temperature. Metal samples were exposed using two oil flow settings giving six temperatures, measured by thermocouples inserted into longitudinal blind holes drilled in specimens. The temperatures are indicated in Fig. 1. Two of the metal temperatures were nominally identical, 820 and 830 °C, but significantly greater corrosion rates were obtained in the run with the greater oil flow (830 °C), possibly because the higher gas velocity would have given greater deposition rates. The ceramic samples were exposed only using the higher oil flow at the three temperatures also shown in Fig. 1. Tests were normally

TABLE 1
The alloys tested and their chemical compositions

Alloy	Fe	Ni	Cr	Al	Ti	Si	Zr	Mn	Y
AISI 310	Bal	20	25	—[c]	—	1.5	—	2	—
Incoloy 800	Bal	35	20	0.3	0.4	0.5	—	—	—
Inconel 600[a]	8	Bal	16	0.3	0.5	0.6	—	0.8	—
50Cr/50Ni	—	50	50	—	—	—	—	—	—
Inconel 589	—	Bal	50	—	—	—	0.9	—	—
Inconel 671	—	Bal	48	—	0.35	—	—	—	—
Kanthal[b]	Bal	—	24	6	—	—	—	—	—
Fe/24Cr (Al, Ti, Y)	Bal	—	24	5.2	0.35	—	—	—	0.05
Fe/24Cr (Al, Si, Ti, Mn)	Bal	—	24	1.4	0.27	1.6	—	0.7	—

[a] Inconel and Incoloy are trade names of International Nickel.
[b] Kanthal is a trade name of Kanthal Steel Co., Sweden.
[c] — No determination.

of 225 h duration with cooling cycles to room temperature after 75 and 150 h.

The ceramics were also exposed at 900°C in melts of sodium sulphate, vanadium pentoxide and an equimolar mixture of the two. A 500 ppm $SO_2/1\%O_2$/balance N_2 gas was bubbled through the melt to simulate boiler conditions. The experimental details have been described previously (1).

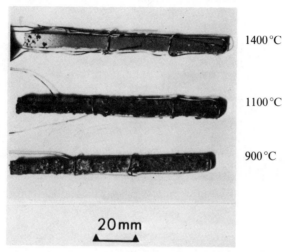

1400°C

1100°C

900°C

20 mm

FIG. 3. Reaction-bonded silicon nitride bars after exposure for 225 h in the burner rig

Exposed samples were photographed and sections removed of typical and extreme corrosion regions for metallographic examination, scanning electron microscopy, energy-dispersive X-ray analysis (EDAX) and measurement of metal loss. The scale from the remainder of the metal rods was removed in a sodium hydride/sodium hydroxide bath and diametral loss measurements made. Typical deposits were analysed chemically and by X-ray diffraction.

The ceramic materials examined were commercially supplied reaction-bonded silicon nitride, hot-pressed silicon nitride and self-bonded silicon carbide, and were exposed as received. The metal alloys and their chemical analyses are listed in Table 1. These alloys were exposed in the as-received condition after grinding to size ($\pm 5\,\mu$m). The 50Cr/50Ni binary and Fe/24Cr-based alloys were made by the CEGB Fabrication Laboratory.

RESULTS

Ceramics

The typical appearance of the ceramic samples after exposure is shown in Fig. 3. All materials formed heavy deposits at 900 and 1100 °C, while at 1400 °C their behaviour was consistent with simple oxidation. Reaction-bonded silicon nitride and self-bonded silicon carbide were both severely attacked, with local attack of up to 2 mm after 225 h at 900 or 1100 °C being

Fɪɢ. 4.   Reaction-bonded silicon nitride exposed in the burner rig for 225 h at 900 °C

measured when the deposits were removed (Fig. 4). The material losses measured for hot-pressed silicon nitride were over an order of magnitude lower (~ 0.02 mm and ~ 0.05 mm after 80 h exposure at 900 and 1100 °C, respectively). The reaction-bonded silicon nitride and self-bonded silicon carbide were also exposed using a fuel containing less than 1 ppm Na and V, when they showed no corrosive attack after 225 h at the three test temperatures.

Diagnostic experiments were carried out using molten salt tests at 900 °C. All the materials were rapidly attacked in sodium sulphate (~ 1 mm h^{-1} for reaction-bonded silicon nitride and self-bonded silicon carbide, and ~ 0.1 mm h^{-1} for hot-pressed silicon nitride). Vanadium pentoxide caused no corrosion after 240 h (Fig. 5), while in an equimolar mixture the corrosion rate for each material was an order of magnitude less than in the pure sulphate.

Metals

The diametral metal losses measured for each alloy after 225 h exposure are

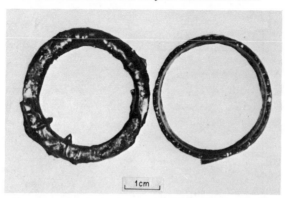

FIG. 5.   Self-bonded silicon carbide tubes after exposure in molten salts at 900 °C. Left V_2O_5, 240 h, right, Na_2SO_4, 5 h

listed in Table 2. On the basis of the morphological examination of the samples, the behaviour of the alloys could be divided into the three groups indicated in Tables 1 and 2.

Fe–Cr–Ni alloys
The data from Table 2 for AISI 310 are plotted in Fig. 6 as a typical example

TABLE 2
Maximum diametral metal losses (mm) measured for each alloy after 225 h exposure

Material	Exposure temperature (°C)					
	720	820	830	960	1 150	1 250
AISI 310	0.1	<0.1	<0.1	1.4	0.8	2.5
Incoloy 800	0.1	0.6	1.5	2.4	0.75	0.6
Inconel 600	0.1	<0.1	4.5	2.7[a]	1.4	0.5
50Cr/50Ni	<0.1	<0.1	0.1	0.5[b]	0.3	<0.1
Inconel 589	<0.1	<0.1	<0.1	2.0[b]	0.2	0.4
Inconel 671	<0.1	<0.1	0.2	0.2[b]	0.2	<0.1
Kanthal	0.1	0.4	—[c]	<0.1	0.4	0.6
Fe/24Cr (Al, Ti, Y)	0.5	<0.1	<0.1	<0.1	<0.1	<0.1
Fe/24Cr (Al, Si, Ti, Mn)	0.2	<0.1	0.2	<0.1	0.1	0.1

[a] 80 h exposure only.   [c] Not determined.
[b] Measured on edges only.

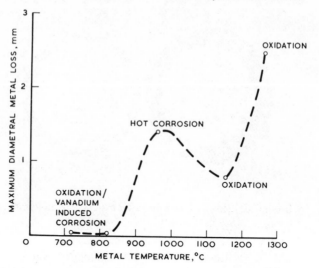

Fɪɢ. 6.   Maximum metal loss after exposure of 225 h in the burner rig as a function
of temperature for AISI 310

of this group. The type of corrosion observed in each temperature range is marked on the figure. Four distinct morphologies can be distinguished:

(1)   720–820 °C. A protective thin chromium-rich scale. At 720 °C some materials showed accelerated attack (type 2). At 820 °C some materials already showed areas of hot corrosion (type 3).

(2)   720 °C. Discrete blisters due to accelerated corrosion. The scale contained vanadium, and discrete chromium-rich sulphides were observed in the underlying metal.

(3)   820–960 °C. Severe oxidation/sulphidation corrosion analogous to that seen on superalloys—'hot corrosion' (Fig. 7).

(4)   1150–1250 °C. Simple oxidation with some spalling, internal oxidation and grain boundary attack.

The morphology of the 'hot corrosion' attack is shown in Fig. 7. The scale consists of a thick iron-rich porous outer oxide, beneath which is a region of oxidation containing iron and chromium oxides and small areas of chromium-rich sulphides. Below this zone, grain boundary sulphidation of the alloy has occurred to form chromium-rich sulphides, totally depleting the remaining metal of chromium and partly of iron. The chromium-rich sulphides have penetrated to a depth of $\sim 300\,\mu$m, beneath which are

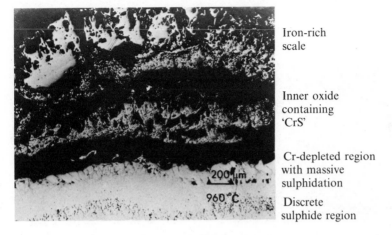

Iron-rich
scale

Inner oxide
containing
'CrS'

Cr-depleted region
with massive
sulphidation

Discrete
sulphide region

FIG. 7. The morphology of hot corrosion on AISI 310 after 225 h in the burner rig
at 960 °C

discrete sulphide particles at grain boundaries associated with considerable porosity.

Cr–Ni alloys

The variations in diametral metal loss with temperature for the various 50Cr/50Ni alloys are listed in Table 2. At all temperatures the samples were covered with a very protective chromium-rich oxide, except that at 960 °C the bottom corners of the rods were severely corroded (Fig. 2). This latter attack showed the typical characteristics of hot corrosion as seen on the Fe–Cr–Ni alloys; viz. a surface layer of chromic oxide (frequently spalled), a layer of chromium and nickel oxides, and a region of internal oxidation.

Figure 8 shows X-ray photographs of the internal oxide region with chromium oxides surrounded by nickel-rich metal. Below this region there are chromium-rich sulphides in a nickel-rich matrix where the chromium content is as low as 3%.

At temperatures above 1000 °C these materials have poor mechanical strength, as is illustrated in Fig. 2(d) where the pressure from the flow of combustion products ($\sim 5\,\mathrm{ms}^{-1}$) was sufficient to bend the samples at 1250 °C.

Fe–Cr alloys

The alloys in this class were based on Fe/24Cr and contained no nickel.

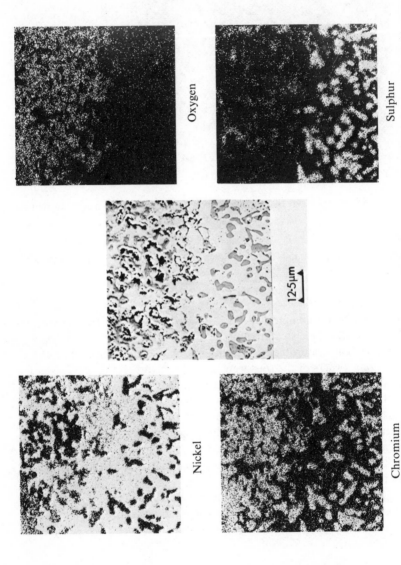

Oxygen

Sulphur

Nickel

Chromium

FIG. 8.   The internal oxide and sulphide region on a 50Cr/50Ni alloy exposed at 960°C for 225 h in the burner rig

12·5μm

These alloys were generally well protected by the formation of an alumina-rich layer (Fig. 2) but at 720 °C (and 820 °C for Kanthal) blister attack occurred, the blisters being Fe–Cr oxides containing vanadium (EDAX). X-ray diffraction analysis of the scale in a blister revealed $(Fe, Cr)_2O_3$ and Na_2SO_4, but no vanadates or vanadium-containing compounds were identified. No oxidation/sulphidation accelerated corrosion was observed at any temperature.

Rapid grain growth of all three alloys occurred at 1150 and 1250 °C, the initial grain size of ~ 0.2 mm having increased to ~ 1–3 mm after 225 h.

DISCUSSION

Ceramics

All three materials showed the same basic behaviour but with different corrosion rates. From the diagnostic melt tests it is evident that the deposition of sodium salts on the ceramic causes the degradation. The maximum temperature for attack is therefore the dewpoint temperature for the condensation of sodium sulphate from the combustion products. For the stoichiometric combustion of distillate oil containing ~ 50 ppm Na this can be calculated to be ~ 1100 °C using the vapour pressure data of Kohl *et al.* (2). The lower temperature of attack is dependent on the mechanism involved. There are two potential mechanisms, both of which can remove the usually protective SiO_2 surface layer:

(1)   The melt on the ceramic is sufficiently thick to depress the oxygen partial pressure at the protective film/melt interface below that required for a protective layer of SiO_2 to be maintained. Volatile SiO is then produced which reacts with the Na_2SO_4 melt to form molten sodium silicates with a free energy of reaction, ΔG, of ~ -84 kcal mol^{-1} at 930 °C (ref. 3). The lowest temperature of severe attack is the melting point of the melt. This would be 880 °C if it were pure sodium sulphate, but vanadium salts depress the melting point and therefore make attack possible at lower temperatures. This mechanism was probably responsible for the catastrophic corrosion rates (Fig. 4) where heavy deposits were present.

(2)   The deposited sodium sulphate reacts with the protective SiO_2 layer to form molten sodium silicates. The lowest temperature of significant attack would be the temperature of the lowest melting point sodium silicate, 780 °C (ref. 4). The sodium silicate with the

lowest melting point for which thermodynamic data is readily available is $Na_2Si_2O_5$, i.e.

$$Na_2SO_4 + 2SiO_2 = Na_2Si_2O_5 + SO_3$$

and examination of the thermodynamics of this equation is instructive. From the data in the Janaf Thermochemical Data Tables (5), the free energy values for this reaction as a function of temperature are:

$T(°C)$	827	927	1027	1127	1227
$\Delta G(\text{kcal mol}^{-1})$	38	32	31	27	23
$\exp\left(-\dfrac{\Delta G}{RT}\right)$	3×10^{-8}	1×10^{-6}	6×10^{-6}	6×10^{-5}	4×10^{-4}

The values of ΔG are inaccurate because of the number of terms in the calculation, and the uncertainties are probably greater than ± 10 kcal. However, they show that the reaction to form $Na_2Si_2O_5$ (and indeed any sodium silicate) becomes more likely at higher temperatures and lower partial pressures of SO_3, for

$$\frac{x_{Na_2Si_2O_5}p(SO_3)}{x_{Na_2SO_4}} = \exp\left(-\frac{\Delta G^T}{RT}\right)$$

where $x_{Na_2Si_2O_5}$ and $x_{Na_2SO_4}$ are the concentrations of $Na_2Si_2O_5$ and Na_2SO_4 and the equation assumes silica is always present. A typical boiler or rig $p(SO_3)$ is 5–10 μbar, so that the thermodynamic data suggest that significant amounts of sodium silicate would be formed at temperatures above 950 °C. These values of $p(SO_3)$ are the non-equilibrium values in the gas; the values could be up to an order of magnitude greater if the surface caused equilibration. Within the available accuracy of the data this degradation mechanism must be considered likely, especially at the higher temperatures and where only limited deposition occurs.

Summarizing the behaviour of each material in detail, the much better corrosion resistance of hot-pressed silicon nitride compared with the other materials is attributed to its greater homogeneity and zero porosity. Self-bonded silicon carbide is attacked preferentially through its free silicon phase, and reaction-bonded silicon nitride is attacked rapidly because its porosity ($\sim 20\%$) means a large surface area is exposed to the deposits. These aspects have been previously discussed in detail (1).

Finally, it should be mentioned that in collaborative work by Burrows (private communication, 1977) samples of reaction-bonded silicon nitride were exposed in the superheater region of an oil-fired power station boiler. Severe corrosion occurred in agreement with the laboratory investigations reported in this chapter.

Metals

It is clear from the results given earlier in the section on Fe–Cr–Ni alloys, that the materials tested suffer from three life-limiting types of corrosion in oil-fired environments. These are as follows:

(1) Accelerated corrosion at 720–820 °C, generally as blisters, due to vanadium/sodium salt deposition. Although this form of corrosion occurs on Fe–Cr–Ni alloys it is more severe on the Fe/24Cr alloys.

(2) Catastrophic 'hot corrosion' in the temperature range 820–960 °C, due to sodium sulphate deposition. The Fe–Cr–Ni alloys are severely corroded, the 50Cr/50Ni alloys suffer severe corrosion only at edges, and the Fe/24Cr alloys are not affected.

(3) High temperature oxidation at 1150–1250 °C. The Fe/24Cr(Al, Ti, Y), Fe/24Cr(Al, Si, Ti) and 50Cr/50Ni alloys are more resistant to oxidation than the other materials.

The susceptibility to corrosion of each material depends on the type of scale which is formed; the latter is determined by the composition of the alloy. Considering first the high temperature oxidation behaviour, the aluminium containing Fe/24Cr alloys forming aluminium-rich oxides give the best protection; the 50Cr/50Ni materials form protective Cr_2O_3 scales but their operating temperature is limited by Cr_2O_3 vaporization; the Fe–Cr–Ni alloys, which form spinel scales, are the least satisfactory. However, the Fe–Cr alloys currently available have very poor mechanical properties. Rapid grain growth makes the alloys brittle and weak. The Fe/24Cr experimental alloys tested had titanium additions to increase the temperature at which rapid grain growth occurred (M. Collins, private communication, 1975) by forming titanium carbides at the grain boundaries. From the present results the temperature for critical grain growth was still below 1150 °C. Poor high temperature mechanical properties also limit the use of the 50Cr/50Ni alloys which melt at 1320 °C.

Secondly, the complex hot corrosion behaviour will be considered. Again the Fe–Cr–Ni alloys were least satisfactory but from Tables 1 and 2 it can be seen that the resistance to hot corrosion of the Fe–Cr–Ni alloys increased as the nickel content decreased and the chromium content increased. As the Fe–Cr alloys containing no nickel were completely resistant it is apparent that nickel is detrimental to hot corrosion resistance. In addition, the

50Ni/50Cr alloys showed hot corrosion only at edges, where the Cr_2O_3 cracked, and the hot corrosion although very severe was less than for the lower chromium content Fe–Cr–Ni alloys. There are several possible explanations of this behaviour:

(1) The initiation time for hot corrosion for Fe–Cr alloys is greater than 225 h. This requires extended work to verify but the morphology observed suggests that it is unlikely.

(2) Aluminium- or silicon-containing scales are most resistant to sodium sulphate induced corrosion. However, hot corrosion studies on gas turbine alloys show that Al_2O_3-forming alloys are less resistant to sodium sulphate induced hot corrosion than alloys forming chromium-rich scales (6).

(3) Nickel, either in the scale or underlying alloy, renders the alloy susceptible to hot corrosion/sulphidation attack. This appears to be the most consistent explanation for the present results.

Hancock (7) has suggested that the formation of a $Ni–Ni_3S_2$ eutectic which is molten above 660 °C is responsible for hot corrosion in sodium sulphate and SO_2-containing environments. However, the morphological examinations in our work showed only chromium sulphides with no evidence of depletion gradients around the sulphides, which would have suggested an exchange reaction between 'CrS' and 'NiS' on cooling. Furthermore, chromium sulphides are more stable than nickel sulphides at the temperatures under consideration. However, the fact is that continuous chromium sulphides did not form in the substrate on Fe–Cr alloys. Although not apparently involved in the propagation stage of the hot corrosion process, it is possible that the initiation step requires the production of a sulphur potential under the scale sufficiently high for a $Ni_3S_2–Ni$ eutectic to form (8). The volume increase associated with the formation of the sulphides cracks the scale reducing the $p(S_2)$ at the interface, so that only Cr sulphides are then stable. The amount of sulphur already injected into the substrate is then so high that continuous chromium sulphides are formed deeper into the metal from the sulphur liberated, as those near the surface are oxidized.

The morphological evidence in this chapter indicates that the formation of sulphides depletes the surrounding metal almost completely of chromium, the depleted metal then oxidizes rapidly as pure nickel or Fe–Ni. The sulphide phase is oxidized to discrete particles of Cr_2O_3 so that a continuous protective oxide phase does not reform.

By analogy with the hot corrosion process in gas turbine alloys it can be presumed that condensation of sodium sulphate is necessary for the attack. As discussed earlier, the maximum temperature for deposition was ~1100°C and therefore for the susceptible alloys this would be the maximum temperature of hot corrosion attack. The lowest temperature at which attack was observed was 820°C, which is below the melting point of sodium sulphate. However, vanadium deposition could give a lower melting point sodium vanadate, and even solid deposits can initiate hot corrosion. The lowest temperature at which attack was observed was material dependent (Table 2), AISI 310 and 50Cr/50Ni alloys showing hot corrosion only above 830°C. This was presumably because of their higher chromium content, but the possibility remains that they would be susceptible to hot corrosion after prolonged exposure at 830°C, since long incubation times for attack are common with many gas turbine alloys. The present 225 h tests have been sufficient to eliminate many materials from consideration, but materials satisfactory over this period need longer term study.

The Fe–Cr alloys were susceptible to an accelerated corrosion process below 800°C, as were to a lesser extent the Fe–Cr–Ni alloys. On both types of alloy the corrosion product was an Fe–Cr oxide containing vanadium. The Fe–Cr alloys investigated normally formed aluminium-rich scales which the vanadium might have fluxed by the formation of vanadates such as $AlVO_4$, which melts at 695°C. The Fe–Cr–Ni alloys showed the same degradation morphology over the whole surface as the Fe–Cr alloys did in blister regions. The corrosion process must therefore be presumed to be the same on both types of alloy once the protective layer has been removed. This mechanism is that which limits oil-fired power station superheater tube temperatures to ~600°C.

CONCLUSIONS

(1) In residual oil-fired environments the deposition of sodium sulphate can cause catastrophic corrosion of silicon-based ceramics due to the formation of molten sodium silicates above ~800°C. Hot-pressed silicon nitride is more resistant than either reaction-bonded silicon nitride or reaction-sintered silicon carbide.

(2) Fe–Cr–Ni alloys should not be used above 700°C in residual oil-fired atmospheres.

(3) Fe/24Cr-based alloys are useful between 800 and 1100°C in residual oil-combustion conditions but at lower temperatures accelerated

vanadium-induced corrosion can occur. Above 1100 °C their mechanical properties are poor.

(4) Of the materials investigated 50Cr/50Ni-based alloys have the highest temperature below which corrosion problems should not occur (850–950 °C). Enhanced corrosion in these materials is found at sharp edges, which should be avoided. Poor mechanical strength limits their use to below 1000 °C.

ACKNOWLEDGEMENT

This work was carried out at the Central Electricity Research Laboratories and is published by permission of the Central Electricity Generating Board.

REFERENCES

1. S. Brooks and D. B. Meadowcroft (1976). 3rd Conf. on Gas Turbine Materials in a Marine Environment, Bath, September (UK Ship Department, Bath).
2. F. J. Kohl, C. A. Stearns and G. C. Fryburg (1975). *Conference on Metal–Slag–Gas Reactions and Processes* (Eds. Z. A. Foroulis and W. W. Smeltzer), The Electrochemical Society, New Jersey, p. 649.
3. D. W. McKee and D. Chatterji (1976). *J. Am. Ceram. Soc.*, **59**, 441.
4. J. Williamson and F. P. Glasser (1966). *Physics Chem. Glasses*, **7**, 127.
5. *Janaf Thermochemical Data Tables* (1968). The Dow Chemical Company, Michigan.
6. D. M. Johnson, D. P. Whittle and J. Stringer (1975). *Corros. Sci.*, **15**, 649.
7. P. Hancock (1961). *First International Congress on Metallic Corrosion*, Butterworths, London, p. 193.
8. V. Vasantasree and M. G. Hocking (1976). *Corros. Sci.*, **16**, 261.

DISCUSSION

A. Nubars (Schoeller Bleckmann Stahlwerke, Kaarst b.Düsseldorf, Germany): Did you also investigate IN-601 (60Ni/23Cr/1.5Al)? We have indications that this alloy behaves fairly well, in particular after pre-oxidation.

Reply: Since the preparation of our paper we have also exposed IN-601 in the burner rig, because this work indicates that with its higher chromium content than IN-600, its corrosion resistance would be better. We found it still suffered severe hot corrosion at 830°C (0.8 mm diametral metal loss) and 960°C (2.4 mm), not quite so catastrophically as IN-600 but no better than AISI-310. At 1250°C its oxidation resistance was superior to IN-600, no measurable metal loss occurring compared with 0.5 mm for IN-600. The effect of pre-oxidation was not investigated but we anticipate that it would not significantly affect our results.

J. Stringer (University of Liverpool): I would like to ask three questions:

(1) You attribute the peak in corrosion rate at 700°C (for example in the 310 ss) to vanadium. Have you considered the possibility that it was due to the well-known alkali iron trisulphate effect?

(2) If indeed vanadium is responsible, why does the effect diminish above 700°C? Vanadium oxides are very stable, and have relatively low vapour pressures, so it is difficult to envisage a mechanism for vanadium corrosion which would diminish at higher temperatures.

(3) Is 50 ppm reasonable for actual systems? Would it not be more economical either to wash the vanadium in the oil to lower levels, or to use an additive such as MgO?

Reply: Answering the questions in turn:

(1) The temperature range investigated was above that where alkali iron trisulphate corrosion has been considered to be important. There was a large concentration of vanadium found in the deposits and it is well known that high vanadium/sodium ratios in a molten salt give very corrosive deposits.

(2) The decreasing corrosion rate above 700°C is a generally observed phenomenon in both vanadium-containing combustion products and in vanadium-free coal-fired environments. No definitive reason can be given at this stage. The solubility of the protective oxide in the melt may decrease with increasing temperature, or the sodium to vanadium deposition ratio may increase with temperature which would reduce the corrosiveness of the deposit. Both aspects are under current investigation at CERL.

(3) 50 ppm sodium and vanadium are typical, indeed in many cases conservative, values of the impurity levels found in residual fuel oil. It is not possible to remove the vanadium from the oil by water washing as it is mainly present as organic compounds, unlike the sodium which is generally present as chloride. Many assessments have been made of the economics and effectiveness of using inhibitors in the fuel or additives such as MgO to control vanadate corrosion in residual oil-fired boilers, and there are many

factors which must be taken into account. In the UK it has been found more economic to operate at 540 °C steam temperature with ferritic superheater tubes to minimize corrosion. Higher temperature components at present tend to be in 50Cr/50Ni-based alloys, which as shown in this paper can be used up to ~1000 °C.

E. Erdoes (Sulzer AG, Winterthur, Switzerland): An additional reason for the limited temperature range of vanadate attack of Fe–Cr–Al alloys, as raised in Professor Stringer's second question, is that aluminium forms aluminium vanadate, which has a temperature range of stability only between ~600 and 800 °C. The formation of this compound might interfere with the corrosion resistance of a Fe–Cr–Al alloy at 720 °C but not at the higher temperatures investigated. On aluminium vanadate there is a paper by W. Richarz and H. Beer in *Thermal Analysis*, vol. 3 (Proc. 3rd ICTA, Davos, 1971), pp. 569–78.

Reply: Thank you. This would be a very good reason why the aluminium-rich oxide was fluxed away only below 800 °C.

A. R. Nicoll (Brown Boveri, Heidelberg, Germany): Concerning the Fe/24Cr alloys containing silicon and aluminium—is the corrosion resistance due to the Cr, Al or Si?

Reply: The primary protection is due to the chromium content giving a chromium-rich scale. At the higher temperatures this is improved by the aluminium and silicon additions, one main reason being that they oxidize internally to form a secondary barrier. However, at 700 °C the presence of the aluminium was probably the cause of the increased corrosion rate. In this environment we consider an alloy with no aluminium or nickel would be most satisfactory.

9

Corrosion–Erosion and Protection Techniques in Furnaces for the Incineration of Household Refuse

A. Moreau

Electricité de France, Paris, France

ABSTRACT

Since their original commissioning, difficulties in the operation of the two main plants for incinerating the household refuse of Paris have been caused by external corrosion of the heat exchanger surfaces. This chapter describes and analyses the corrosion–erosion phenomena and separates them into different mechanisms. The analysis of the corrosion–erosion rates shows that they are dependent on six main factors.

Once these factors have been established the corrosion–erosion rates are correlated with each of them using the data in a principal component statistical analysis. The data were the incidence of bursting and the ultrasonic thickness measurements of the generator tube bundles as a function of their time of operation.

A description is given of the modifications and of the protection techniques examined and put into operation to reduce the corrosion–erosion rates— these included various tube steels and protective metallic coatings. Finally, a new type of superheater platen protected by a thin layer of highly conducting refractory resistant to corrosion and erosion is described in detail and the operational experience obtained with it is discussed.

INTRODUCTION

Steam-raising by incineration of household refuse (typical composition given in Table 1) poses considerable problems in a great many installations; one of these is the corrosion–erosion of the heat exchanger surfaces.

The City of Paris's Service for the Industrial Processing of Urban Refuse comprises three installations for the incineration of 1 600 000 tons per year.

A. Moreau

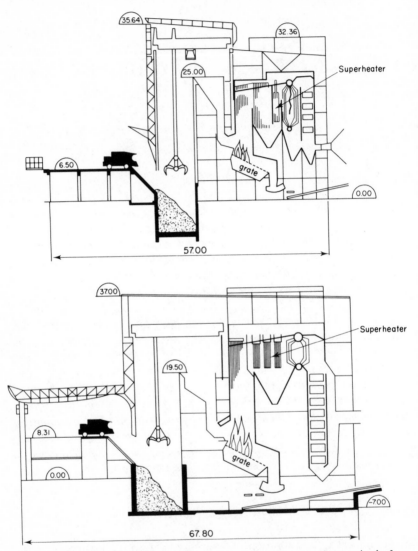

FIG. 1. Schematic diagrams showing the locations of the superheaters in the Issy-les-Moulineaux (top) and Ivry (bottom) plants (all dimensions in metres)

TABLE 1
Typical composition of household garbage in the Paris area

Materials	(%)
fine solids < 8 mm	10.35
fine solids, 8–19 mm	11.75
vegetable and putrescible matter	16.46
paper	32.18
metal	4.14
rags	2.88
glass	10.76
bone	0.97
unclassified combustible debris	3.05
unclassified non-combustible debris	2.4
plastic	5.06
Breakdown of plastic matter	
polyvinylchloride	23.89
polyethylene	49.14
polystyrene	14.48
various	12.49
Breakdown of metals	
scrap iron	88.95
non-ferrous	11.05

Of these, the two most modern, which are at Issy-les-Moulineaux and Ivry-sur-Seine (Fig. 1) are considerably affected by corrosion and erosion. Before modification, the lifespan of the tubes in the critical zones was approximately 6000 h.

The Issy-les-Moulineaux works has four boilers generating superheated steam at 64 bars and 410 °C. The incineration capacity of each furnace is 19 tons per hour.

The Ivry-sur-Seine works has two boilers generating superheated steam at 96 bars and 470 °C. Their incineration capacity is 50 tons per hour each.

Immediately on commissioning these two units, operational difficulties were experienced with external corrosion of the heat exchanger surfaces. One can differentiate between two sorts of corrosion:

(1) Corrosion of the Radiantly Heated Walls
The corrosion occurs in a reducing medium. In spite of a heavy excess of air, 12 % O_2, certain zones are in an atmosphere which is alternately oxidizing and reducing. The heterogeneity of the fuel and the rapid variations in the heat release rates are the causes of this. This type of corrosion is also

TABLE 2
Approximate composition of the ash deposits

Elementary analysis	(%)
iron	19.95
silicon	1.78
aluminium	1.7
chromium	1.33
lead	6.7
sodium	3.31
potassium	8.37
calcium	1.6
magnesium	1.68
zinc	5.12
chloride	4.23
total sulphur	8.83
phosphorus	0.25
tin	0.18
manganese	0.12
copper	1.18
carbon	0.46
oxygen and non-titrated material	34.21
Constitution	
ferric chloride	5.75
ferric oxide	26.36
alumina	3.21
silica	3.81
chromium oxide	1.94
lead oxide	7.22
sodium sulphate	10.21
potasium sulphate	18.65
zinc sulphate	12.63
lime	2.24
magnesia	2.78
phosphoric anhydride	0.57
oxide of tin	0.23
oxide of manganese	0.15
oxide of copper	0.22
carbon	0.46

water and non-titrated material to make 100

| *pH value of the aqueous solution* | 4.5 |

FIG. 2.   Heavy fouling of the superheater tubes (1/5 scale)

encountered in coal-burning boilers. Our problem has been solved satisfactorily by protecting the walls with a heat-conducting SiC concrete.

(2) Corrosion of the Superheaters

It is this type of corrosion and its remedies which are the subject of this study. For its consideration the following factors must be taken into account:

(a)   the composition of the tube steel (Chromesco III, $2\frac{1}{4}$Cr 1Mo);

(b)   the heavy coverage of the tubes with ash (Fig. 2)—note the high degree of fouling during incineration;

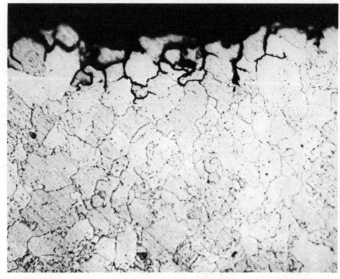

FIG. 3.   Metallographic section showing corrosion attack on the tubes ($\times$ 500)

(c)   the composition of the ash, which contains a considerable number of components, particularly sulphates (Table 2); and

(d)   the type of corrosion (Fig. 3).

Micrographs of a superheater tube show a general intergranular attack with high nodular penetration.

Reaction mechanisms are badly defined in the complex deposits produced by refuse incineration and a great many different explanations have been given.

For this reason, we chose a different approach to the problem, using the observations as a basis for defining the factors which cause corrosion and determining the importance of the various factors.

CHOICE OF RELEVANT EXPERIMENTAL VARIABLES

The variable factors chosen should be such that their magnitudes will allow the corrosion–erosion rates to be predicted as closely as possible. For this reason, we used two approaches for the choice of these factors:

(1)   Taking account of the factors which by experience are known to be important in this type of phenomenon: temperature of the metal and temperature of the combustion gases.

(2)   Taking account of the following factors suggested by observations made during operation.

First Observation

We observed very great differences in the corrosion rates which were inexplicable by the usual factors. Consequently, we sought to use these differences to find new explanatory factors. For this, we were able to benefit from the differences in the construction of the superheaters at the two plants (Fig. 4). In both cases, the superheaters are of the pendant type but, at Issy-les-Moulineaux, the tubes are parallel to the gas flow, except in zones C and F, while at Ivry-sur-Seine all the tubes are perpendicular to the combustion gas flows.

In Fig. 5 we have attempted to show the layout and to summarize the operational conditions in the two plants, with special emphasis on three points:

(1)   Corrosion–erosion rate is high when the ash-laden gases flow perpendicularly to the tubes.
(2)   Corrosion–erosion rate is low when the ash-laden gases flow parallel to the tubes.
(3)   Corrosion–erosion rate is low when, because of some local factor (in this case, the tube hanger), the ashes remain stuck to the tubes.

Second Observation

A modification made in 1973, which resulted in the acceleration of the corrosion–erosion rate in the Ivry-sur-Seine BT superheater by a factor of 10, brought to light the importance of the speed of the ash particles carried in the combustion gases; this was independent of the speed of the carrying gases.

Observation of the jet of particles revealed considerable differences in speed (4 and 13 m/s), dependent on whether the upstream tube bank had a narrow or a broad interspace (Fig. 6).

After the modification, the impact energy of the particles was increased in the same proportion as the increase in corrosion–erosion rate. The speeds of the particles were measured from the tracks recorded on photographic plates.

The repeated impacts on the tubes of a closely spaced bank produced a reduction of the speeds of the particles relative to those of the carrying gases. There is a critical tube spacing at which the braking effect becomes significant, so this phenomenon does not occur when the upstream tube bank has a broad interspace.

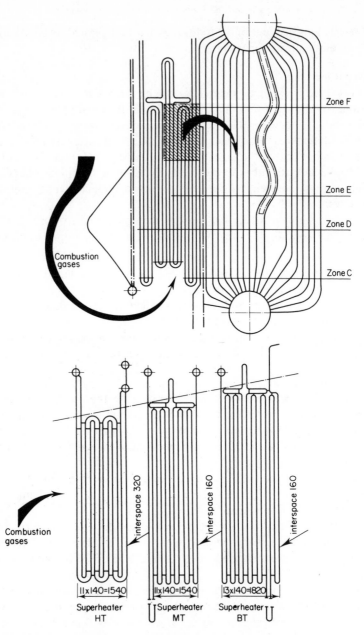

FIG. 4. Diagram of the superheater configurations and the gas flow directions in the two plants; (top: Issy-les-Moulineaux, bottom: Ivry)

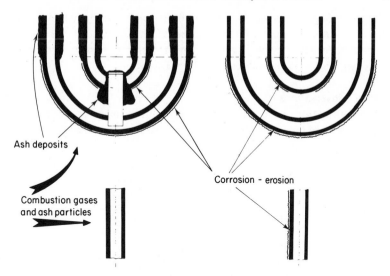

Fig. 5. Corrosion zones and ash deposition on the various tube configurations

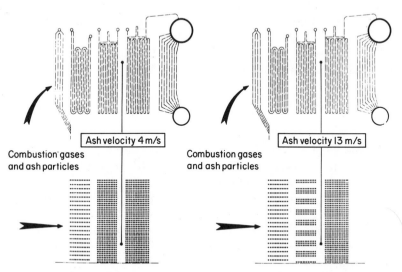

Fig. 6. Effect of modification of the tube spacing of the MT superheater in the Ivry plant on the ash velocity in the BT superheater. Top: side elevation; bottom: plan

The two observations brought to light a new explanatory factor for the corrosion–erosion which we have termed VCV, which is V sin α, where α is the angle between the velocity vector of the particles and the centreline of the tubes, and V is the velocity of the ash particles.

Further Observations
Other observations have led us to take account of the inverse of the ceiling distance, the lateral positioning, and the distance from the soot blower.

TABLE 3
Summary of the critical factors for corrosion–erosion rates (COR)

TFM	temperature of the gas
TVP	steam temperature
VCV	speed of the ashes × sin α
Y	inverse of the ceiling distance
POS	lateral position
R	distance from the soot blower

Without going into detail regarding the measuring methods used for quantifying each of the factors in Table 3, it can be stated that the measurement of the ash velocity was by chronophotography and that of the corrosion–erosion rates by ultrasonic thickness measurements. We have taken several tens of thousands of measurements since 1965 in these tube banks during each overhaul or repair of the boilers.

STATISTICAL STUDY

On the one hand we have the determining factors listed in Table 3 and, on the other hand, the corrosion–erosion rates (COR) of the individual tubes which we draw up as a data table. In order to use this table of data we employ a method for representing and classifying the individuals in one plane, known as principal component analysis. This method makes it possible (Fig. 7) to represent the assembly of individual data points and to show their grouping together into clusters, and to characterize the clusters thus brought to light.

On this projection (Fig. 7), the various tube banks are clearly separated, with the exception of the BT superheater which is split into two clusters

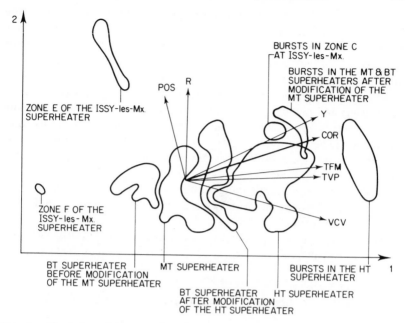

FIG. 7. Clustering of the corrosion–erosion rate and bursting data in the plane of the first two principal components (1 and 2)

representing the data points obtained before and after the modification of the MT superheater.

The three superheater assemblies, HT, MT and BT, after the modification to the MT superheater, form high corrosion–erosion rate clusters. with the MT and BT superheaters inverted in position relative to their true position. Their relative positions and this inversion already provide an indication of the small part played by temperatures in this range of values.

The BT superheater, before the modification was made to the MT superheater and the E and F zones of the Issy-les-Moulineaux superheaters (Fig. 4), formed low corrosion–erosion rate clusters.

The individual data points in the clusters of tube bursts (corresponding to corrosion–erosion rate maxima) tend to be offset towards the line of increasing corrosion rate (COR on Fig. 7), and this may be a genuine effect of the position parameters, POS and R, in the boiler.

Bursts in the elbow-bends of zone C of the Issy-les-Moulineaux superheater are closely grouped together on account of the homogeneity of their appearance times. The corresponding corrosion–erosion rates are considerably higher than the rates in the right-hand parts of zone E, where the tubes are parallel to the gas flow. The difference in rates between these two thermally similar zones is remarkable. This disparity proves the effectiveness of the 'protection' which is afforded by the orientation of the tube axes parallel to the ash and gas flow direction.

Principal component analysis provides an interesting representation of the clusters and highlights a number of factors particularly likely to cause corrosion–erosion: the orientation of the tubes relative to the gas-flow direction; the bombardment by the flying ashes; and the temperature of the gas and of the steam, which were found to be well correlated with the corrosion–erosion rates.

To make progress in explaining the corrosion–erosion, we must attempt to distinguish between factors which are correlated fortuitously and those which are truly dependent. Of course, no statistical technique allows us to do this for sure but to explore the data as exhaustively as possible we have supplemented the principal component analysis by an analysis of variance.

Analysis of Variance
We will merely set out here the conclusions of this analysis.

Gas temperature and metal temperature effects
It is concluded that when the temperature of the gases is greater than 650 °C and that of the metal is greater than 290 °C, the corrosion–erosion rate will be independent of these two temperatures. Thus, the effect of temperature saturates at these two levels.

Lateral positioning effect
The corrosion rate is slightly greater at the centre of the tube banks than at the sides. The effect of distance from the centre is small but sufficient to increase the probability of bursting at the centre of the tube bank.

Velocity of the flying ashes and orientation effect
Above the saturation temperatures, this effect is significant and preponderant. It does not become slight until the velocities are below 4 m/s.

CONCLUSIONS OF THE STATISTICAL STUDY

Analysis of the data summarized briefly above shows that the corrosion rates of the superheater tubes are not proportional to the gas and metal temperatures when the latter are greater than 650 °C and 290 °C, respectively. Further, we deduced that:

(1)   The first two rows, which are more exposed to the impact of the particles, are always more corroded than the following ones.

(2)   The increase in gas passage cross-section at the MT superheater, which increases the speed of the particles upstream of the BT superheater from 4 to 13 m/s, causes rapid destruction of this tube bank.

(3)   The right-hand parts of the Issy-les-Moulineaux superheater which lie parallel to the flow of the gas and ashes are practically free from corrosion–erosion in contrast to the situation for the perpendicular lower elbow-bends.

Consequently, the only parameters which influence corrosion–erosion at temperatures higher than the saturation levels are the orientation of the tubes and the speed of the ashes.

For the incineration of household refuse at temperatures greater than 650 °C for the gas and 290 °C for the metal of the tubes, the techniques for combating corrosion–erosion of the superheaters are thus:

(a)   to lay out the pipes parallel to the gas flow direction;

(b)   when perpendicular tubes are used small closely spaced barriers should be inserted to break the speed of the ash particles, thus protecting the downstream parts of the boiler;

(c)   very low speeds, or ample dimensioning to limit the speeds of the ash particles to 4 m/s; and

(d)   protection of the tubes, which is the objective of the second part of this chapter.

PROTECTION OF THE TUBES

Before describing the eventual solution, the tests carried out are listed for reference only:

Grades of steel
Of all the steels tested, a 25Cr/20Ni steel, although it did not eliminate the corrosion–erosion entirely, reduced the rate by half relative to the Chromesco III originally installed. Co-extruded tubes, such as 50Cr/50Ni on Incoloy 800, did not give any better results.

Metallization
Many different metallization techniques were tried but without useful results. Strontium zirconate coatings were not tested because of their prohibitive cost.

Enamelling
This type of protection was not degraded, but its fragility forced us to abandon it.

Chromizing
No results.

Injection of magnesium compounds
This injection was abandoned because of lack of convincing results.

Faced with these failures, our efforts concentrated on the placing of the superheaters in panels, and the protection of these panels by ceramic layers. The layer must have the following properties:

(a) the constituent material must have a high thermal conductivity;
(b) it must be chemically inert when exposed to the combustion gases, the ashes and to condensed salts;
(c) it must be possible to install it as a fine coat (6 mm) and it must adhere perfectly to the metal surface of the tubes and studs (Fig. 8); and
(d) it must be easy to place and in particular must be cold-hardening to allow handling.

Ceramic Bonding Material, with Hydraulic Cold-Hardening
Composition:

SiC	84.4
Kaolinite	6.0
Ca aluminate (CA, CA$_2$)	6.3
Borax	0.6
Alumina	2.2
	99.5%

This material has a high SiC content. The Ca aluminate allows it to harden cold and the presence of borax slightly delays the hardening mechanism, which facilitates placing on the panel. The clay should ensure the hardening above a certain critical temperature, with ceramic bonding.

FIG. 8. Coating of the panel-type superheaters with ceramic layers

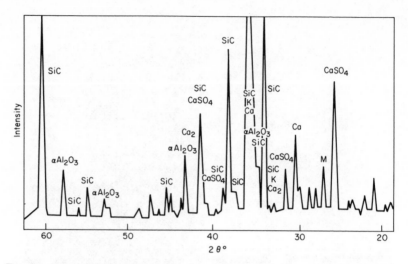

FIG. 9. X-ray diffractometer trace of ceramic-bonded silicon carbide after three months operation, with identification of the peaks (M, montmorillonite; K, kaolinite; SiC, silicon carbide; CA, CaO, Al_2O_3; CA_2, CaO, $2Al_2O_3$; $CaSO_4$, calcium sulphate)

However, after three months of operation of the boiler, we observed by X-ray diffraction studies (Fig. 9) and energy-dispersive X-ray analyses (Fig. 10) of the mass of the product, the presence of a kaolinite type clay, CA/CA_2-type binders, and also of $CaSO_4$.

The presence of $CaSO_4$, confirmed by the infrared absorption spectra (Fig. 11), allows us to affirm that reaction has occurred between the calcium aluminate hydraulic binder and the combustion products.

Figure 12 gives the evidence for the formation of deposition products containing sodium, potassium, iron, titanium, lead and zinc.

Chemically Bonded Ceramic
Composition:

SiC	80
Al_2O_3	7
Clay	8
$Al(H_2PO_4)_3$	5
	100 %

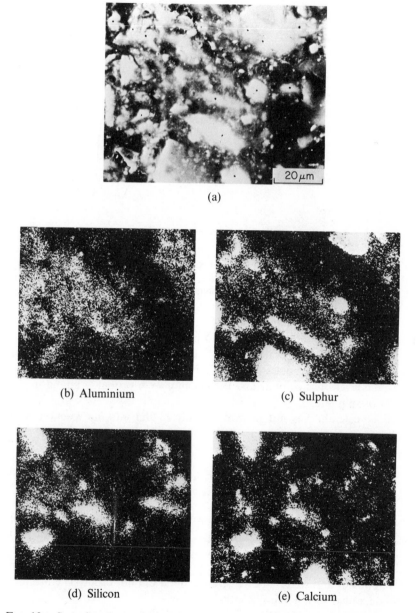

(a)

(b) Aluminium   (c) Sulphur

(d) Silicon   (e) Calcium

FIG. 10. Scanning electron micrograph (a) and elemental X-ray images (b–e) of ceramic-bonded silicon carbide after three months operation

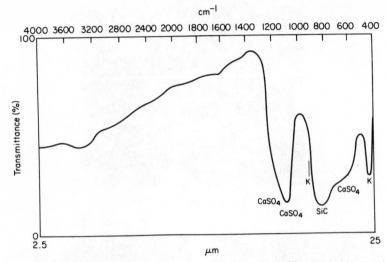

FIG. 11.   Infrared absorption spectra of ceramic-bonded silicon carbide showing presence of calcium sulphate (K, kaolinite; SiC, silicon carbide; CaSO₄, calcium sulphate)

This material, which has a high SiC content, is chemically bonded by the reaction between the aluminium acid phosphate and the Al_2O_3.

After a period of operation, we noted (Figs. 13 and 14) the presence of large quantities of potassium, sulphur and calcium in the pores of the ceramic layer. The sulphur and the calcium were probably bound in the form of sulphate.

Since calcium was not present initially in the mix, its reaction with sulphur in the combustion products should not degrade the bond.

CONCLUSIONS ON THE USE OF CERAMIC LAYERS FOR PROTECTION

As we have already seen, the ceramic layer with a hydraulic binder was not adequate. Indeed, the combustion gas containing calcium chloride and sulphur tended to react with the calcium of the binder and to destroy it; further, since the service temperature was 600–700 °C maximum, development and sintering of a ceramic bond did not occur. The product was degraded within a few months.

On the other hand, the layer with a chemical bonding agent was satisfactory for two reasons:

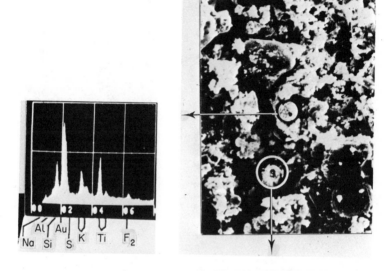

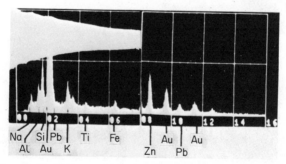

FIG. 12.   Structure and energy dispersive X-ray analysis of deposits on ceramic-bonded silicon carbide after three months operation

A. Moreau

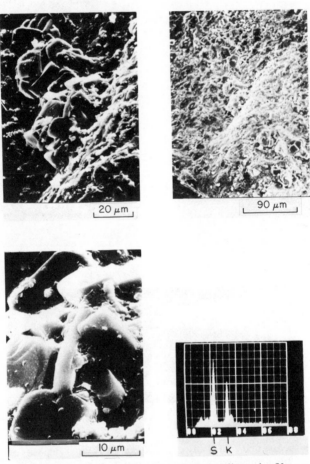

FIG. 13.   Scanning electron micrographs and energy-dispersive X-ray analysis of fracture surface of chemically bonded silicon carbide after operation

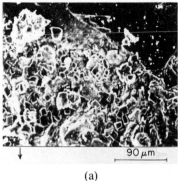

(a)

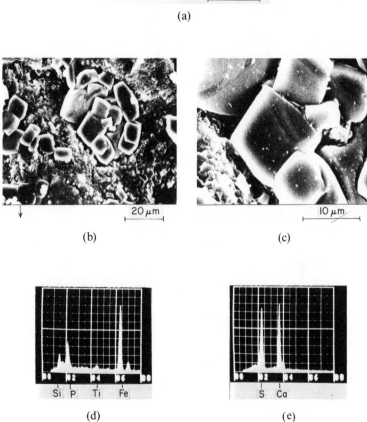

(b)   (c)

(d)   (e)

FIG. 14.   Scanning electron micrographs (a, b, c) and energy-dispersive X-ray analyses of small (d) and large (e) crystals on the fracture surface of chemically bonded silicon carbide after operation

(1)   The metal was attacked slightly by the phosphoric acid, which provided extremely good adherence and good heat transfer.

(2)   When combustion products condensed on the surface of the product or even inside the open porosity they did not react with the chemical bond.

A part of the Ivry-sur-Seine superheaters has, since January 1975, been formed of panels protected in accordance with this technique; since that time, no important degradation or failure has occurred in this zone.

10

Corrosion Behaviour of Hot-pressed and Reaction-bonded Silicon Nitride in Condensed Phases

E. Erdoes and H. Altorfer

Sulzer Brothers Ltd, Winterthur, Switzerland

ABSTRACT

With a view to the utilization of silicon nitride in gas turbines its corrosion behaviour in molten salts has been investigated by crucible tests. The salt melts were based on sodium sulphate with additions of sodium chloride, sodium chloride + vanadium pentoxide, sodium chloride + lithium oxide and also on the eutectic melt of sodium, magnesium and calcium sulphate. Testing temperature was 900°C. The gas phase above the melt was air, artificial combustion gas with 0.2% SO_2 or pure argon. Additional experiments were run in a test rig at 950°C with the addition of sodium, sodium + magnesium and of sodium + vanadium (at 850°C).

The tests were interpreted by scanning electron microscopy, X-ray elemental analysis and X-ray diffraction analysis of the corrosion products, and by gravimetry. It was established that sodium sulphate produces a local attack on hot-pressed silicon nitride with the formation of a foamy corrosion product, which shows the diffraction pattern of cristobalite with a small amount of tridymite. With reaction-bonded material the attack can destroy the sample completely.

In acid and oxidizing conditions the corrosion is mitigated, while in an alkaline environment it is enhanced. In the presence of magnesium sulphate and calcium sulphate, forsterite develops on the surface of the silicon nitride.

Rig testing confirmed the results from the crucible tests. With the addition of magnesium we observed the formation of a crystalline phase of the osumilite type and the corrosion rate was reduced.

Our investigation led to the conclusion that the corrosion resistance of silicon nitride in the presence of sulphates is highly dependent upon the environment. Three cases can be distinguished:

	Environment	Corrosion resistance
(1)	*Oxidizing and acidic*	*good to excellent*
(2)	*Oxidising and mildly alkaline*	*fair*
(3)	*Reducing and mildly alkaline*	*bad.*

Highly alkaline conditions are not relevant for gas turbines and an acidic and reducing environment is not realizable if we consider the equation

$$SO_4^{2-} - O^{2-} \rightarrow SO_3 \rightarrow SO_2 + \tfrac{1}{2}O_2.$$

It should be emphasized that various crystalline phases could be formed on the surface of silicon nitride by the corrosion reactions. These could be of technical interest for minimizing corrosion by the use of coatings and additives.

INTRODUCTION

Silicon nitride, a ceramic material containing very little oxygen, is being closely considered for application in gas turbines (1). Its stability against oxidation is quite satisfactory; appreciable oxidation has been noticed only above 1000 °C for reaction-bonded material and above 1300 °C for hot-pressed material (2).

Several studies have been made of the behaviour of silicon nitride in gases (3) but information on the influence of molten salts is rather scarce. Schlichting (4) has discussed the hot corrosion of silicon carbide and silicon nitride by molten salts. The aim of this chapter is to contribute to our knowledge of the morphology of the corrosion products of silicon nitride exposed in melts which could be found in gas turbines.

The main analytical method has been scanning electron microscopy supported by X-ray diffraction analyses for phase identification. Other methods have been used when needed.

Silicon nitride is commercially available in different grades: reaction-bonded (porous) or hot-pressed (dense) material. In the reaction-bonded material, α-Si_3N_4 is the main constituent and the hot-pressed material consists principally of β-Si_3N_4 with up to 5 % magnesium oxide. From the investigations of Jack *et al.* α-Si_3N_4 is known to be oxygen bearing (5). Both α- and β-Si_3N_4 have a hexagonal lattice but they differ in their space groups and lattice constants.

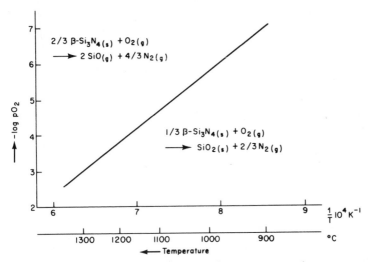

FIG. 1. Oxidation of β-Si$_3$N$_4$ as a function of temperature and oxygen partial pressure

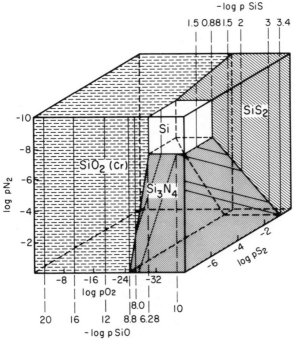

FIG. 2. Phases in the system Si–O–N–S at 1200 K (Si$_2$N$_2$O has not been considered)

E. Erdoes and H. Altorfer

TABLE 1

Free enthalpy of chemical reactions in the system Si–O–N–S at 1200 K (reactions with S_2 and O_2)

Reaction	ΔG(cal/mol)	log Kp	
$\frac{1}{3}Si_3N_{4_s} + O_{2_g} \rightarrow SiO_{2_s} + \frac{2}{3}N_{2_g}$	$-138\,610$	$\log p^{\frac{2}{3}}N_2 \cdot p^{-1}O_2 =$	25.24
$\frac{2}{3}Si_3N_{4_s} + O_{2_g} \rightarrow 2SiO_g + \frac{4}{3}N_{2_g}$	$42\,028$	$\log p^2 SiO \cdot p^{\frac{4}{3}}N_2 \cdot p^{-1}O_2 =$	7.655
$\frac{1}{3}Si_3N_{4_s} + S_{2_g} \rightarrow SiS_{2_s} + \frac{2}{3}N_{2_g}$	-4404	$\log p^{\frac{2}{3}}N_2 \cdot p^{-1}S_2 =$	0.802
$\frac{2}{3}Si_3N_{4_s} + S_{2_g} \rightarrow 2SiS_g + \frac{4}{3}N_{2_g}$	$+32\,910$	$\log p^{\frac{4}{3}}N_2 \cdot p^2 SiS \cdot p^{-1}S_2 =$	-5.99
$2SiO_g + O_{2_g} \rightarrow 2SiO_{2_s}$	$-235\,191$	$\log p^{-2}SiO \cdot p^{-1}O_2 =$	42.83
$2SiO_g + S_{2_g} \rightarrow 2SiS_g + O_{2_g}$	$+74\,938$	$\log p^2 SiS \cdot pO_2 \cdot p^{-1}S_2 \cdot p^{-2}SiO =$	-13.65
$SiO_{2_s} + S_{2_g} \rightarrow SiS_{2_s} + O_{2_g}$	$+134\,206$	$\log pO_2 \cdot p^{-1}S_2 =$	-24.44
$2SiO_{2_s} + S_{2_g} \rightarrow 2SiS_g + 2O_{2_g}$	$+310\,129$	$\log p^2 SiS \cdot p^2 O_2 \cdot p^{-1}S_2 =$	-56.48
$Si_s + O_{2_g} \rightarrow SiO_{2_s}$	$-166\,276$	$\log p^{-1}O_2 =$	30.28
$2Si_s + O_{2_g} \rightarrow 2SiO_g$	$-97\,361$	$\log p^2 SiO \cdot p^{-1}O_2 =$	17.73
$\frac{3}{2}Si_s + N_{2_g} \rightarrow \frac{1}{2}Si_3N_{4_s}$	$-41\,500$	$\log p^{-1}N_2 =$	7.56
$Si_s + SiO_{2_s} \rightarrow 2SiO_g$	$+68\,915$	$\log p^2 SiO =$	-12.55
$Si_s + S_{2_g} \rightarrow SiS_{2_s}$	$-32\,070$	$\log p^{-1}S_2 =$	5.84
$2Si_s + S_{2_g} \rightarrow 2SiS_g$	$-22\,423$	$\log p^2 SiS \cdot p^{-1}S_2 =$	4.084
$Si_s + SiS_{2_s} \rightarrow 2SiS_g$	$+9647$	$\log p^2 SiS =$	-1.757
$2SiS_g + S_{2_g} \rightarrow 2SiS_{2_s}$	$-41\,717$	$\log p^{-1}S_2 \cdot p^{-2}SiS =$	7.60
$\frac{4}{3}Si_3N_4(\beta)_s + O_{2_g} \rightarrow 2Si_2N_2O + \frac{2}{3}N_{2_g}$	$-145\,467$	$\log p^{2/3}N_2 \cdot p^{-1}O_2 =$	26.49
$\frac{1}{3}Si_3N_{4_s} + SiO_{2_s} \rightarrow 2SiO_g + \frac{2}{3}N_{2_g}$	$+96\,581$	$\log p^2 SiO \cdot p^{2/3}N_2 =$	-17.59

Reference: O. Knacke and I. Barin (1973). *Thermochemical Properties of Inorganic Substances*, Springer Verlag, Heidelberg.

There are two different modes of oxidation of silicon nitride:

(a) $\qquad \frac{2}{3}Si_3N_4(s) + O_2(g) \rightarrow 2SiO(g) + \frac{4}{3}N_2(g)$

(b) $\qquad \frac{1}{3}Si_3N_4(s) + O_2(g) \rightarrow SiO_2(s) + \frac{2}{3}N_2(g)$

In the first case we have 'active' oxidation, with the formation of gaseous reaction products; in the second case 'passive' oxidation, with the formation of a protective layer. Figure 1 gives the limits of the active and passive oxidation modes as a function of oxygen partial pressure and temperature.

Sulphur is known to play an important role in the process of corrosion of metallic materials by combustion gases. To assess the influence of sulphur in the formation of corrosion products of silicon nitride we calculated the stability diagram for different oxygen and sulphur partial pressures taking into account Si, Si_3N_4, SiO_2 (cristobalite), SiS_2, SiO and SiS. Si_2N_2O was disregarded (Fig. 2). The formation of silicon sulphide is predicted to be unlikely for corrosion in combustion gases in accordance with experience. The basic equations and numerical values for the calculations are listed in Table 1.

EXPERIMENTAL TECHNIQUES, RESULTS AND DISCUSSION

We began our experimental work by exposing different commercial grades of hot-pressed and reaction-bonded silicon nitride in two different salt melts:

(1)   52.5 mol % Na_2SO_4; 39.6 mol % $MgSO_4$; 6.9 mol % $CaSO_4$; 1.0 mol % NaCl.

(2)   99 mol % Na_2SO_4; 1 mol % NaCl.

Samples (0.4–0.9 g) were washed with distilled water, dried at 110 °C and then immersed in Al_2O_3 crucibles filled with melt for 1 week at 900 °C. After exposing the samples we washed them thoroughly, dissolving all soluble matter, and dried the residue again. Before and after the exposure the sample was analysed by X-ray diffraction. The results show the marked differences between the attack in the two different melts.

The sodium sulphate/sodium chloride melt destroyed the reaction-bonded samples completely and only powdery residues could be found. Although the predominant phase originally had been α-Si_3N_4, now the residue was enriched in β-Si_3N_4. The hot-pressed samples were only slightly attacked. The alkaline earth sulphate melt attacked all samples to an intermediate extent.

After these preliminary experiments further crucible tests were carried out with hot-pressed Si_3N_4, using similar experimental techniques.

Condition of surface: ground to 600 grade.
Qualitative analysis: Si, traces of Fe and W.
Phase analysis: β-Si_3N_4, trace of α-Si_3N_4, small amounts of WSi$_2$.
Composition of melt: 99 mol% Na$_2$SO$_4$ + 1 mol% NaCl, air, 900 °C/168 h.

This melt acts as an etchant (Fig. 3) and produces a partly blistered surface structure. These blisters could be explained by the formation of gaseous nitrogen according to the reaction:

$$Si_3N_4 + 3O_2 \rightarrow 3SiO_2 \text{ (cristobalite)} + 2N_2, \Delta G^\circ_{1200} = -415.8 \text{ kcal}$$

Cristobalite and tridymite were detected. Figure 4 shows the blisters in transverse section.

The attack on the surface is not only dependent on the composition of the melt (sodium sulphate/sodium chloride) but also on the atmosphere above it. If we changed the air for a gas mixture of 19% O_2, 2% CO_2, 0.2% SO_2, remainder N_2 we found with the same melt (though only after 50 h) neither corrosion nor etching (Fig. 5).

The attack of sodium sulphate and the protective action of sulphur dioxide can be understood from the equation:

$$SiO_2 + Na_2SO_4 \rightleftarrows Na_2SiO_3 + SO_2 + \tfrac{1}{2}O_2, \Delta G^\circ_{1200} = +33.9 \text{ kcal}$$

The effect of an addition of vanadium pentoxide to the melt is interesting and was studied in the following composition: 98 mol% Na$_2$SO$_4$ + 1 mol% NaCl + 1 mol% V$_2$O$_5$, air, 900 °C/168 h. The immersion in this melt leads to the formation of a crystalline surface layer (Fig. 6). These crystals were analysed and they contained aluminium (!) and sodium besides silicon. By X-ray diffraction we established the presence of a primitive cubic structure with $a_0 = 9.086$ Å, in addition to α-cristobalite and tridymite. We assume the formation of nosean, $Na_8Al_6(SiO_4)_6SO_4$, space group P43m through reaction with the Al_2O_3 crucible. The formation of nosean crystals is a good example of the ability of silicon nitride to form all sorts of crystalline layers *in situ*; we encountered this again and again.

Deposits in industrial gas turbines almost invariably contain alkaline

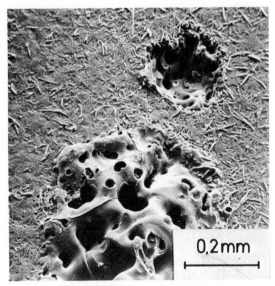

FIG. 3.   Corrosion of hot-pressed Si_3N_4 in sodium sulphate/sodium chloride melt at 900°C

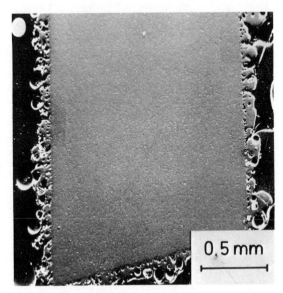

FIG. 4.   Same as Fig. 3—transverse section

FIG. 5.   Inhibition of Na_2SO_4 attack by SO_2

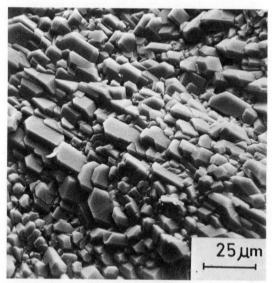

FIG. 6.   Crystalline layer on hot-pressed Si_3N_4 after exposure to sulphatic melt
with V_2O_5 addition (Al_2O_3 crucible)

earth sulphates. Having found a decisive influence of such sulphates in the preliminary tests we proceeded to a more thorough investigation in a melt of composition: $52 \, mol \%$ $Na_2SO_4 + 39.6 \, mol \%$ $MgSO_4 + 6.9 \, mol \%$ $CaSO_4 + 1 \, mol \%$ NaCl, air, $900 \, °C/168 \, h$.

After this exposure the sample was covered with a mat of fine crystals (Fig. 7) containing magnesium. X-ray diffraction analysis shows the

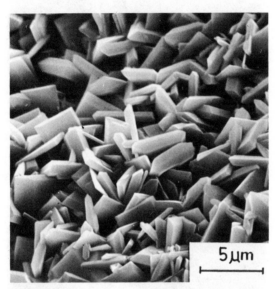

FIG. 7. Crystalline layer of forsterite and enstatite on hot-pressed Si_3N_4

presence of forsterite, Mg_2SiO_4, and of minor amounts of enstatite, $MgSiO_3$. Here we find a reaction with magnesium sulphate according to

$$2MgSO_4 + SiO_2 \rightarrow Mg_2SiO_4 + 2SO_2 + O_2, \Delta G^\circ_{1200} = +11.0 \, kcal$$

Because of the frequent addition of up to 5% MgO, hot-pressed silicon nitride contains forsterite and enstatite as well as glassy phases (6). MgO has an adverse effect on oxidation resistance (2), but the resistance to molten salt corrosion is increased.

Melts containing free alkalis attack silicon nitride markedly, as shown by tests in a melt of composition: $98 \, mol \%$ $Na_2SO_4 + 1 \, mol \%$ NaCl $+ 1 \, mol \%$ Li_2O, air, $900 \, °C/168 \, h$. The surface is corroded (Fig. 8) and again we find the formation of well-shaped crystals (Fig. 9).

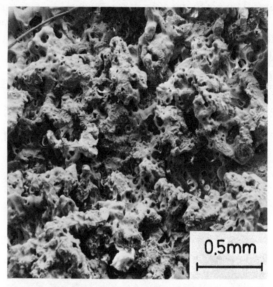

FIG. 8. Corrosion of hot-pressed Si_3N_4 by addition of Li_2O to sulphatic melt

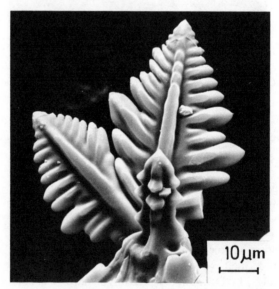

FIG. 9. Same as Fig. 8—detail

TABLE 2

Type of material	Temperature (°C)	Time (h)	Additions (ppm)
Hot-pressed	950	200	30Na
Reaction-bonded	950	200	30Na
Hot-pressed	950	200	30Na + 15Mg
Reaction-bonded	950	200	30Na + 15Mg
Hot-pressed	850	300	15Na + 100V

In addition to the crucible tests we exposed samples of silicon nitride in an oil-fired combustion rig under the conditions given in Table 2. All additions were as oil-soluble naphthenates. The fuel was a light oil with 0.4 % sulphur. The corrosion results are in general agreement with the crucible experiments, but there is no qualitative difference between the hot-pressed and reaction-bonded material.

Figure 10 (hot-pressed Si_3N_4, 30 ppm Na, 950 °C/200 h) is representative of the type of attack. With the addition of 30 ppm Na + 15 ppm Mg (hot-pressed Si_3N_4, 950 °C/200 h) we found, in addition to the blistered glassy surface, the appearance of another crystalline phase

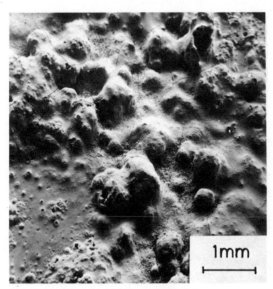

FIG. 10.   Corrosion in combustion rig exposure (30 ppm Na)

FIG. 11.   Crystal formation in combustion rig exposure (30 ppm Na + 15 ppm Mg)

FIG. 12.   Inhibition of corrosion of hot-pressed Si_3N_4 by V_2O_5 in combustion rig

(Fig. 11) which contains magnesium and a trace of iron. It was determined by X-ray diffraction that a new phase had been formed with hexagonal lattice a = 10.15 Å, c = 14.21 Å, c/a = 1.400 and space group P6cc or higher symmetry.

These lattice constants are in close agreement with those of the mineral osumilite (K, Na, Ca) (Mg, Fe^{II})$_2$(Al, Fe^{II}, Fe^{III})$_3$(Si, Al)$_{12}$O$_{30}$ with a = 10.17 Å, c = 14.34 Å, c/a = 1.410, space group P6/mcc. We do not propose that osumilite had been formed but a compound having a similar crystal structure.

Samples of hot-pressed Si_3N_4 were exposed in the eutectic (Na$_2$, Mg, Ca) SO$_4$ melt and formed a layer of forsterite and enstatite. In the subsequent exposure in the combustion rig this layer proved not to be stable and osumilite-like crystals developed.

In the presence of vanadium pentoxide (15 ppm Na + 100 ppm V) the attack on silicon nitride was markedly diminished (Fig. 12) and the oxide film consisted mainly of cristobalite (this test was at 850 °C for 300 h).

Quantitative measurements on the hot-pressed silicon nitride materials exposed in the combustion rig tests gave the following results:

Experimental conditions	Weight loss/area
950 °C/200 h/30 ppm Na	17 mg/cm^2
950 °C/200 h/30 ppm Na + 15 ppm Mg	11 mg/cm^2
850 °C/300 h/15 ppm Na + 100 ppm V	5 mg/cm^2

It has to be borne in mind that under *completely non-oxidizing conditions* silicon nitride is unstable in sulphatic melts. A sample of hot-pressed material could be dissolved in a melt of 99 mol % Na$_2$SO$_4$ + 1 mol % NaCl with purified argon above the melt at 900 °C.

CONCLUSIONS

To sum up the results we reach the following conclusions:

(1) The stability of silicon nitride in sulphatic melts is a function of the acidity (alkalinity) of the melt and the oxidizing (reducing) conditions of the melt and the atmosphere above it.

(2) In oxidizing and acidic conditions the corrosion resistance of silicon nitride is excellent.

(3) In oxidizing and neutral to alkaline conditions the corrosion depends on the alkalinity of the melt. In alkaline and reducing conditions silicon nitride tends to become unstable.

(4)   Reducing and acidic conditions are not likely to be encountered at 900 °C for

$$SO_4^{2-} - O^{2-} \rightarrow SO_3 \rightarrow SO_2 + \tfrac{1}{2}O_2$$

REFERENCES

1. D. E. Stoddart (1972). *Gas Turb. Int.* (July–August), 16.
2. A. Fickel (1973). *CZ Chem. Tech.*, **H4** S1.
3. R. Ebi (1973). Diss., Univ. Karlsruhe; and S. C. Singhal (1974). In *Ceramics for High Performance Applications* (Eds. J. J. Burke, A. E. Gorum and R. N. Katz), Brook Hill Pub. Co., Massachusetts, p. 533.
4. J. Schlichting (1975). *Werkstoffe Corros.*, **26**, 753.
5. I. Colquhoun, S. Wild, R. Grieveson and K. H. Jack (1973). *Proc. Brit. Ceram. Soc.*, **22**, 207–28.
6. D. W. Richerson (1973). *Ceram. Bull.*, **52**, 560.

DISCUSSION

J. A. Klostermann (Technische Hochschule, Eindhoven, Netherlands): You have shown that silicon nitride crystallizes easily, forming different crystalline phases with impurities. Do you have any indications that the mechanical behaviour is influenced by the crystallization? The crystals could initiate a fracture.

Reply: In our work we have confined ourselves to the corrosion phenomena of silicon nitride. Mechanical behaviour under the influence of corrosion has not been studied by us.

M. Betz (MTU, München, Germany): Concerning the influence of corrosion on the mechanical properties, I can add that in MTU experience ceramics parts become damaged after service in hot gases and that this can lead to a deterioration in the mechanical properties.

A. Naoumidis (Kernforschungsanlage, Jülich, Germany): MgO is added to Si_3N_4 for the fabrication of hot-pressed parts. Does this not have an adverse effect on the corrosion behaviour of Si_3N_4? What would be the effect of SO_2 or SO_3 in this case?

Reply: We have no indication of such an effect, which should, however, be studied by using samples with different MgO additions.

11

Application of the Acoustic Emission Technique to the Detection of Oxide Layer Cracking During the Oxidation Process

C. Coddet, G. Béranger and J. F. Chrétien
Université de Technologie de Compiègne, France

ABSTRACT

The acoustic emission produced during the deformation of constrained materials is already used in numerous laboratories to study crack initiation and crack propagation mechanisms.

The application of this method to the observation in situ *of the cracking of oxide layers during their growth at high temperature has required the development of special techniques; effectively, the high temperature of the oxidation process is incompatible with the use of the actual piezoelectric transducers. Thus, it has been necessary to use a waveguide between the transducer located outside of the furnace and the sample itself. Moreover, the exact position of the emission source has to be located with care in order to avoid ambiguous signals. Therefore a system of linear localization of the source with the help of two transducers has been developed.*

The first results obtained have clearly shown cracking of titanium oxide layers and have provided evidence on subsequent changes undergone by the sample during cooling.

INTRODUCTION

During the growth of oxide films at high temperature, certain stresses will appear. The relief of these stresses may occur principally in two ways: creep or cracking. It is the second process we are going to investigate.

Frequently, the cracks in oxide layers are observed micrographically after cooling of the samples. This cooling to room temperature induces

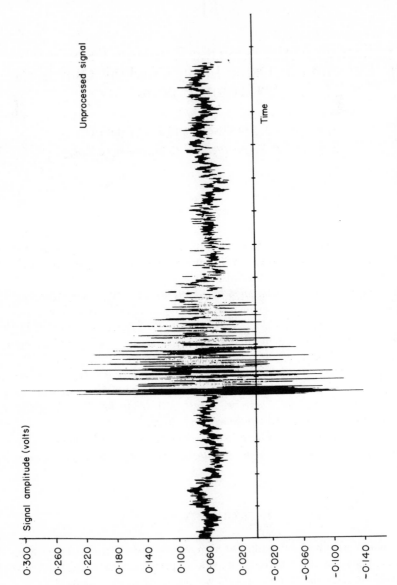

FIG. 1. Typical acoustic emission signal recorded during crack propagation

thermal stresses which may produce new crack initiation and propagation in the oxide or the metal.

Thus, it seems necessary to detect *in situ* the formation of cracks during the growth of oxide layers. Experimental methods which allow this study are few and difficult; one can mention the vibrational technique proposed by Bruce and Hancock (1) and the electrical resistivity measurements. We therefore attempted to develop a new highly sensitive technique: the acoustic emission method.

Irreversible damage in a solid produces elastic waves. The detection and analysis of these waves may allow the study of a cracking mechanism. Generally, the elastic waves produced (still called 'acoustic emission') are typical of the irreversible process by which they are generated; for example, crack initiation or propagation (2) (Fig. 1).

EXPERIMENTAL TECHNIQUE

Acoustic emission signals are usually detected by high sensitivity unbacked piezoelectric ceramic transducers. But, the emission level is very low and the frequency range very wide; consequently detection is difficult.

As a general rule the transducer is coupled to the sample by an oil film. The sensor output passes to a charge amplifier followed by an amplifier, a high pass filter and a signal analysis system (Fig. 2).

Due to the complexity of the collected signals at this stage, we only recorded a continuous voltage proportional to the number of acoustic

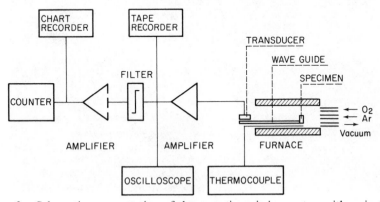

FIG. 2. Schematic representation of the acoustic emission system with a single transducer

emission events higher than a preset threshold in order to eliminate the background. The count rate or total count vs. time are displayed on a XY graphic recorder.

In the high temperature range, where oxide growth takes place, transducers are not readily available. Thus we have used waveguides to ensure the transmission of the acoustic emission from the sample, at the furnace temperature, to the transducer at room temperature. The first experiments were conducted with stainless steel waveguides. Later, we expect to use alumina guides which seem more appropriate.

FIRST RESULTS

The first tests have been carried out with pure titanium at 800 °C. This material gives an oxide layer with numerous cracks parallel to the metal/oxide interface (Fig. 3) (refs. 3 and 4).

The acoustic emission total count vs. time curve for a sample shows a first stage during which no acoustic emission was detected, followed by intense

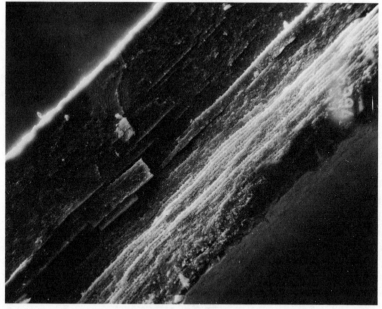

FIG. 3.   Scanning electron micrograph of the oxide scale formed on titanium in pure oxygen at 800 °C for 72 h (× 350)

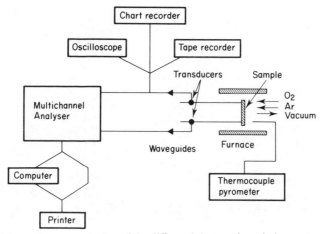

FIG. 4. Schematic representation of the differential acoustic emission system with two transducers

emission periods separated by times with no noticeable acoustic activity. It seems that these observations could easily be related to the numerous and successive accelerations of the oxidation rate recorded on the kinetic curves (5). So, we have tried to locate the emission source using a differential acoustic emission monitoring system (Fig. 4) with two transducers. This device, by measuring the difference in arrival time of the acoustic emission

FIG. 5. General view of the acoustic emission device—see also Fig. 6

FIG. 6.   Further view of the acoustic emission device

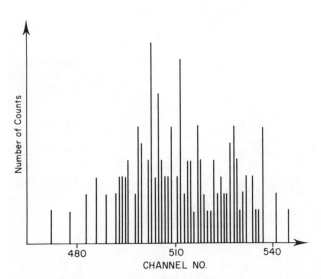

FIG. 7.   Number of events vs. channel no. recorded during the cooling of the sample after isothermal oxidation

signals to the two sensors, allows the location of the source. The positions of the acoustic emission sources are displayed on an oscilloscope screen (Figs. 5 and 6).

However, this system shows a great lack of sensitivity in comparison with the counting technique, and at the present time we have only succeeded in recording the intense acoustic emission which occurs during the slow cooling of the sample after the oxidation test (Fig. 7). The intensity of this emission indicates that the structure of cooled oxide layers observed with a microscope is probably different from that existing at high temperature.

CONCLUSION

Our first experimental results have shown that the acoustic emission method is very efficient for the continuous study of a high temperature oxidation process to detect structural changes. The first results obtained corroborate the hypothesis of oxide layer fracture (and possibly of the oxygen saturated metal beneath the oxide layer) during the high temperature oxidation reaction of titanium with oxygen. The sensitivity of the acoustic emission monitoring system has still to be improved to locate the emission source during the oxidation process, but the analysis of the strong signals emitted during the cooling of the sample clearly showed numerous structural changes. Therefore it appears necessary in oxidation studies to perform *in situ* experiments in order to reach a correct knowledge of the oxidation mechanism.

REFERENCES

1. D. Bruce and P. Hancock (1969). *J. Inst. Metals.*, **97**, 140.
2. J. F. Chrétien and N. Chrétien (1972). *Non-destruct. Test*, **8**, 220.
3. P. Sarrazin and C. Coddet (1974). *Corros. Sci.*, **14**, 83.
4. C. Coddet, J. F. Chrétien and G. Béranger (1976). *C.R. Acad. Sci. Paris*, **t282**, sér. C, 815.
5. X. Lucas, E. A. Garcia, G. Béranger and P. Lacombe (1975). *Proc. 25th Internat. Mtg Société de Chimie Physique* (Dijon, July 1974), Elsevier, New York.

DISCUSSION

R. Pichoir (ONERA, F-92320, Chatillon, France): Could you state the maximum operating temperature of your apparatus?

Reply: The maximum operating temperature is in fact only limited by the nature of the waveguide material. In our first experiments, performed at 800 °C, stainless steel waveguides were used; in further experiments, at higher temperature, the use of alumina waveguides is foreseen.

D. B. Meadowcroft (CERL, Leatherhead, UK): (1) Which stainless steel is used for the waveguides? As it will also oxidize and oxide spalling occurs on cooling would the fact that this was not detected give a lower limit to the sensitivity for the thickness of film which can be studied. (2) How is the waveguide connected to the specimen?

Reply: (1) The stainless steel used for the waveguides was a 18Cr/10Ni steel. In our experimental conditions, only a very thin oxide film was formed on the waveguide and no spalling process occurred. (2) The sample was held to the waveguide by a screw.

R. C. Hurst (Euratom, Petten, The Netherlands): I would like to ask three questions. The first is, do you notice any noise from the waveguide itself due to bending at high temperatures? The second is, can you test specimens under stress to see any resultant oxide fracture? The final question concerns the sensitivity of the technique in that titanium oxidizes very rapidly and such large oxide failures should obviously be detectable. However in practice, other alloys should oxidize much less and therefore have you examined any other alloys to really test the viability of the technique?

Reply: (1) Several tests were performed without a sample; in this case no noticeable emission was detected; we concluded that the stainless steel waveguides were appropriate in our experimental conditions.

(2) In this work no external stress was applied to the sample, except that due to thermal cycling.

(3) The first steps in our research were intended to test the proposed method. Therefore, we chose first to study the oxide growth on titanium, because the oxide scale was known to exhibit intensive and typical cracking during its growth. Further studies will be devoted to other alloys, particularly refractory alloys.

P. Kunzmann (Schweiz. Aluminium AG, Neuhausen, Switzerland): Can you analyse the frequency spectrum of the acoustic emission?

Reply: The analysis of the frequency spectrum is a very complex problem and has not yet been achieved.

M. Betz (MTU, München, Germany): Can you explain the effect given in the last figure, showing that many channels have the same number of counts?

Reply: In Fig. 7 of our paper the number of events recorded for each channel is very low due to the large number of channels. Thus, the number of events recorded in each channel of the recorder may be the same.

R. Hales (BNL, Berkeley, UK): 18Cr–8 % Ni alloys have a reversible α–γ transformation which breaks the Cr_2O_3 scales on them. This would give a measure of rig sensitivity.

12

Profile of Requirements for Protective Coatings Against High Temperature Corrosion and Some Experience with their Behaviour in Service

W. BETZ

Motoren- und Turbinen-Union, München, West Germany

ABSTRACT

A successful use of protective coatings depends upon weighing the advantages and disadvantages of the coatings against their costs. The assessment has to be made for every intended application with consideration of the specific loadings on turbine blade and coating.

Three groups of loadings are described: thermal loading, mechanical loading and chemical loading. From these, some requirements for coatings have been derived.

A purely synthetic way of assessing coating performance or a purely empirical assessment in service does not seem viable at present. MTU is testing coatings using conditions for the most important parameters closely similar to those in service, i.e. creep tests in a hot, corrosive gas stream and load–temperature cyclic tests with cooled blades in a hot, corrosive gas stream.

One result is that some parts with coatings have been shown to have a shorter lifetime in certain stress and temperature ranges than uncoated parts.

In spite of the well-known difficulties, an approach is outlined for combining various data measured on coatings and on parent material to assess coatings for specific applications.

INTRODUCTION

The question whether protective coatings should or should not be used in aero engines is primarily a matter of the costs involved. The cost-saving

185

potential of protective coatings is based mainly upon two expectations:

(1) Lengthening of the lifetime of a component in a corrosive service atmosphere until it is comparable with the lifetime in an inert atmosphere.

(2) Preservation of the surface integrity of the component, so that turbine efficiency is maintained for as long a service period as possible.

However, these potential gains must be weighed against the costs and losses involved, namely: costs of producing the coating—these costs may arise repeatedly during the lifetime of the component; and shortening of the lifetime of a component through damage when the coating is being produced or when in service.

TABLE 1
Weighing the usefulness of protective coatings against high temperature corrosion

Gain	Loss
Longer lifetime through less attack by corrosion	Costs of producing the coating, perhaps repeatedly
Efficiency maintained through better surface integrity	Shorter lifetime through damage to parent material when coating is applied

The gains therefore have to be weighed against the expenditure (Table 1). A separate assessment of the usefulness of protective coatings must be made for each combination of engine, protective coating and type of service involved and will certainly differ appreciably between stationary gas turbines and aero engines.

It is necessary in all cases to obtain a picture of the lifetime of the parent material/coating combination under service-like conditions, as measured data have shown that this is by no means always identical with the lifetime of the coating in the case of corrosion attack alone.

This is shown as a function of stress in Fig. 1, in which the stress in the test is plotted against the time a component takes to rupture. In a completely inert atmosphere, line A is the breaking point of the material. This is its maximum attainable lifetime. In corrosive combustion gas, this line shifts to shorter times to rupture (B), depending on the aggressiveness of the corrosion. Our measurements have shown that under relatively high stress,

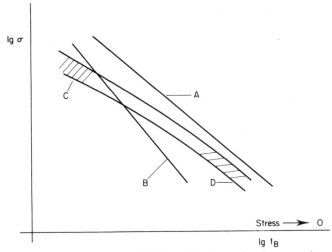

FIG. 1. Schematic dependence of lifetime (t_B) of coated and uncoated components on operating environment and stress (σ). A: Inert atmosphere without protective coating. B: Corrosive atmosphere without protective coating. C: Corrosive atmosphere with protective coating; high stress level. D: Corrosive atmosphere with protective coating

i.e. giving a time to rupture of up to 500 h, certain protective coatings cause a drop in the lifetime of the parent material/protective coating combination (C). On the other hand, measurements of weight loss and penetration depth show that, when the stress tends to zero, coated test specimens are superior to uncoated specimens but are nevertheless destroyed eventually after consumption of the coating: the consequence is thus a zone D, which in conjunction with C reflects the actual behaviour of coated components.

For a given target lifetime the mechanical and corrosion damage to be expected can be assessed. It can then be ascertained whether a particular protective coating/parent material combination will provide a longer or a shorter lifetime (in comparison with the unprotected component). Even in the latter case the maintenance of efficiency owing to the better surface integrity may offset the higher production costs.

SERVICE LOADINGS

The long-term loading is not the only factor affecting the lifetime of the component. Basically, the lifetime limits are determined, on the one hand, by the properties of the coating/parent material combination itself (they are regulated during the production process, for instance by the heat treatments involved) and, on the other hand, by the imposed service loadings.

For aero engines, the service loadings can be summarized and roughly quantified as follows, there being a certain latitude between engines used for military and civil purposes:

The thermal loading (Table 2) comprises combustion gas temperatures up to 1400 °C and material temperatures up to 1050 °C. During starting-up

TABLE 2
Thermal loading on operational gas turbine blades

Type of loading	Magnitude	Effect
Temperature range:		Diffusion processes
in combustion gas	up to 1 400 °C	
in material	up to 1 050 °C	Changes in structure
local	up to 200 °C/mm	
temperature gradients		Mechanical stresses
time-wise	up to 100 °C/s	

and shutting-down of the engine, temperature gradients both in time and space are set up owing to the cooling of the components. These are of the order of 100 °C/s and 200 °C/mm, respectively. They result in diffusion processes, changes in structure and mechanical stresses.

Mechanical loadings (Table 3) result from centrifugal stresses in the 170 N/mm^2 range in the zone where there is a risk of rupture in the case of rotor blades. Local stress gradients (cyclic loading), e.g. through temperature changes and variations in speed, are the cause of the extremely complex overall stress distribution. This results in cyclic strain of the component, and such strain may lead very quickly to the formation of cracks in the coating. These cracks may travel into the parent material and substantially shorten the fatigue life.

As a result of the high gas velocity, there is also a loading through erosion, foreign object impact and the breakaway of products of corrosion.

According to the mechanical and thermal loadings, the components have design lifetimes ranging from ~1000 h to 5000–10 000 h.

Finally, the chemical loading (Table 4) is caused by oxidation in the excess air of the combustion gas (~12 % O_2 by volume), by contamination

TABLE 3

Mechanical loading on operational gas turbine blades

Type of loading	Magnitude	Effect
Centrifugal stress	$\sim 170\,N/mm^2$	Cyclic strain in all temperature ranges
Local	e.g.	
} stress gradients	$\pm 30\,N/mm^2$	Formation of cracks in the coating
Time-wise (difference in temperature, gas pressure)		
Gas velocity	up to 600 m/s	Stripping of the coating, erosion, mechanical removal
Foreign object impact		Spalling of the coating

T, σ: lifetimes from 1 000 h to 5 000–10 000 h.

in the fuel (max. permissible sulphur 0.3 % by weight, usually 0.01 % by weight), as well as by corrosive constituents of the intake air (depending on flight conditions, e.g. sea salt up to 1 ppm or industrial atmosphere). Chemical attack can be affected by the pressure (up to 25 bar) and the flow rate (up to 600 m/s) of the combustion gases. Chemical attack results in destruction of the surface of the unprotected material and local penetration of corrosion especially along grain boundaries, combined with changes in the alloy composition in contiguous zones. Roughening of the surface causes a drop in the efficiency of the engine.

TABLE 4

Chemical loading on operational gas turbine blades

Type of loading	Magnitude	Effect
Excess oxygen	$\sim 12\,\%$ by volume	Oxidation and corrosion as depletion across surface or local attack
Contamination in the fuel, e.g. sulphur	0.3 w/o permissible 0.01 w/o usual	Alloy impoverishment
Contamination of the intake air, e.g. sea salt, industrial atmosphere	up to ~ 1 ppm	Roughening of the surfaces
Pressure	up to 25 bar	
Flow rate	up to 600 m/s	

REQUIREMENTS OF PROTECTIVE COATINGS

From the service loadings and from the basic requirement for economic prolongation of operational life, a series of requirements for protective coatings can be derived (Table 5):

(1)   When they are produced, the parent material should not be changed in significant areas, e.g. at grain boundaries, either thermally or chemically.

(2)   There should be little interaction with the parent material during the design lifetime, so that the mechanical properties of the load-bearing material are not impaired.

(3)   Low susceptibility to cracking is required, i.e. there must be adequate ductility or the ductile/brittle transition temperature must be low enough.

(4)   The viscosity of the coating must be high enough to prevent it being stripped by the centrifugal force.

(5)   The coating must have sufficient corrosion resistance to prevent it being used up prematurely during the design lifetime.

(6)   Uniform, smooth depletion to preserve best possible surface integrity and to avoid local damage in depth, is also desirable.

(7)   The production cost of the coating must be within acceptable limits.

In order to assess a specific combination of coating and parent material, these requirements should be checked, one after another, to find out whether they have been met. For the assessment and weighting of the individual items, the problem is that their individual contribution to the damage processes and their interaction with one another are often only inadequately known.

TABLE 5
Basic requirements for protective coatings

Economic prolongation of functioning

Individual requirements which may serve this purpose:
    little damage to the parent material when the coating is being produced or in service
    little proneness to cracking, i.e. ductility down to lowest possible temperatures
    adequate viscosity
    little and uniform, smooth depletion of the coating
    acceptable production cost

The alternative method of making an integral assessment based on service experience in the test engine is ruled out, owing to the time and cost involved.

TEST FACILITIES AND RESULTS

At MTU, an endeavour is therefore made to assess protective coatings by means of laboratory tests under as service-like conditions as possible, i.e. we try to simulate those service parameters which are important for service life and check the similarity with service conditions by comparison with results from test engines. A brief account of some of the results is given below.

Figure 2 shows the time to rupture for coated and uncoated turbine blades as a function of the test stress. The components are heated by a

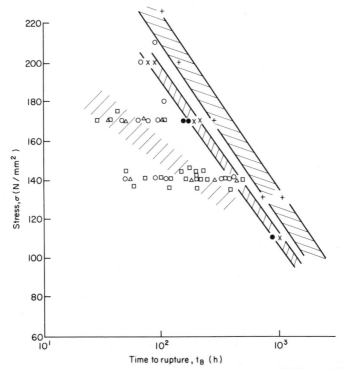

FIG. 2. Creep behaviour of IN-100 in hot gas stream at 950 °C. + Flat test specimens in air. Flat test specimens in hot gas: × uncoated; ● Cr-aluminized. Turbine blades in hot gas stream: ○ coated; □ pack aluminized; △ Cr-aluminized

stream of combustion gas to 950 °C and have a uniform temperature distribution across the section where the rupture is expected. The times to rupture for uncoated parts in hot gas lie in a relatively narrow scatter band below that obtained in air for the IN-100 material used in this case. If the blades are aluminized by a variety of processes, the times to rupture come down, in some cases substantially, and are more scattered than for

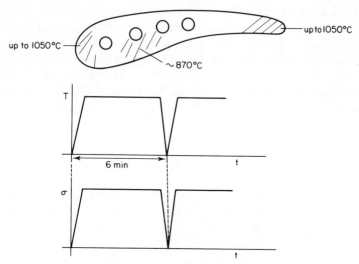

FIG. 3.   Loadings in a test to simulate service conditions

uncoated blades. So far, one coating process from the series of pack processes (in the wider sense) has proved the most worth while. This process, chrome aluminizing, provides coatings which involve no significant drop in life in the stress range for lifetimes up to ∼ 1000 h. In this case, the lifetimes of the uncoated test specimens are reached in the range tested and a much better surface quality is maintained.

This can be interpreted as indicating that, both in the coated and in the uncoated condition in this case, the mechanical loading of the parent material mainly determines the lifetime.

In order to improve still further the similarity of our tests to service conditions, we have combined the following types of loadings which are important in flying operations in another test rig (Fig. 3):

(1)   High velocity combustion gas from aviation fuel as corrosive medium.

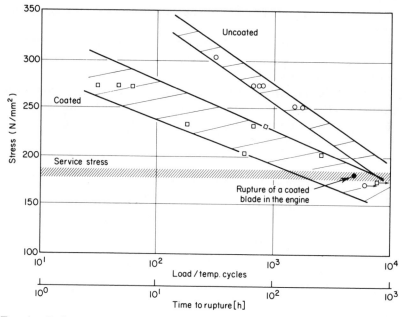

Fig. 4. Fatigue stress/time-to-rupture curves for coated and uncoated blade materials

(2) Realistic temperature distribution in the component through internal cooling.

(3) Cyclic temperature change from room temperature up to service maximum temperatures of ~1000°C at the hottest sites of the component in ~20 s followed, after a holding time of just under 6 min, by just as rapid cooling to room temperature.

(4) Cyclic stressing synchronously with the temperature changes, with maximum values during the holding time, leading to rupture in a test time of acceptably long duration.

Figure 4 gives the results as time to rupture or number of cycles as a function of the stress. Qualitatively, the same situation obtains as in the creep test: uncoated components have a longer lifetime under high stress than aluminized blades. In this case, too, the chromium/aluminide coating provides the exception—it does not shorten the lifetime.

To assess how close the test is to service conditions, the service stress on an engine blade in the maximum stress range is shown. Coated blades from this engine reached the times shown under test before distinct cracks were formed. There seems to be quite good agreement between test and service conditions.

In this test, too, there is an indication that when lifetimes are quite long, i.e. stress is correspondingly lower, the two curves can be expected to approach or cross each other.

The next illustration (Fig. 5) can be used to interpret this situation and to point out at the same time a way in which it may be possible to incorporate

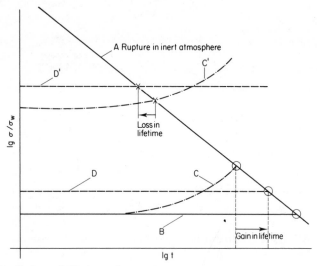

FIG. 5.   Prediction of lifetime of coated components under stress in corrosive
atmospheres

weight loss, penetration depth and similar isolated measured data into the concept of considering the lifetime of a component as the major criterion for assessing protective coatings.

The ratio of the effective stress σ to some reference stress for the material σ_w, is plotted against the logarithm of service time elapsed. (A typical reference stress, σ_w, might be taken as the fatigue stress for 100 h endurance at 0.1 cycles/h.) Line A shows the time to rupture of the uncoated component in a completely inert atmosphere and without changes in the structure caused by corrosion. For these conditions, line B plots the ratio σ/σ_w while a component is in service. If, on the other hand, there is corrosion, the effective stress goes up in the course of time through material consumption over the surface and, still more, through damage at grain boundaries (notch effect), and curve C represents the σ/σ_w ratio for the component.

If the component is given a coating, the material may be damaged as a result of the required treatment, σ_w decreases and, if the coating resists all corrosion, curve D now represents the σ/σ_w ratio for the component. Despite the damage to the material, the coating has provided a longer lifetime.

However, if the process described is shifted to a higher stress level, the picture is reversed: the curve for corrosion C becomes the curve C′, essentially maintaining its shape, and the drop in strength from line B to D applies equally for D′, so that the coated component ruptures earlier than the uncoated one.

An attempt can be made to show other loadings of the coating/parent material combination in this diagram in an analogous way. Admittedly, practical application to the analytical prediction of lifetime requires comprehensive knowledge of the individual processes and of their interaction, if any. Such a use of this method of presentation is therefore not yet in sight and we are still having recourse to component testing under service-like conditions.

DISCUSSION

U. W. Hildebrandt (BBC-Forschung, Heidelberg, Germany): I would like to refer to the question of the use of such coatings. If one considers the times to rupture lines, these lines do not intersect before a time which is nearly the lifetime of the aircraft engine. From this point of view a coating would not be necessary. On the other hand, the figures of the analytical method you showed were not very convincing. What we finally want is an increase in the lifetime and not an aesthetic analytical method.

If for reasons of improvement in efficiency you want a smoothing of the surface why do you not apply a wear-resistant coating containing hard phases?

Reply: The reason is that the roughening of surfaces of uncoated blades is due to corrosion and this means that you would have to find wear-resistant coatings which are at the same time corrosion resistant,

Ch. Just (Sulzer AG, Winterthur, Switzerland): In rotating blades one must also take into account the effect of the centrifugal forces on the lifetime. All coatings have a low strength at high temperature so that the mass of the coating must be carried completely by the base metal. The higher stress due to these factors can reduce the lifetime remarkably. This is true in particular for thin section hollow blades or thick coatings.

Reply: This is a significant factor and in very bad cases there might be an increase of stress between 5 and 10%.

Granacher (Technische Hochschule, Darmstadt, Germany): Why is the rupture strength of gas turbine blades in combustion atmospheres lower than that of flat specimens?

Reply: We suppose that this mainly depends on the different casting parameters used for these parts. Another reason might be that in the turbine blade there is not such an ideal uniaxial stress as in the flat specimen.

P. Felix (BBC, Baden, Switzerland): Creep rupture tests of different coated alloys revealed that the reduction in the time to rupture is mainly caused by the heat treatment during the coating process. Therefore, it can be assumed that this damage can be removed in many alloys by a supplementary solution heat treatment.

Reply: This is true, but is only part of the truth. There are coatings which produce a greater loss in lifetime of the part than does the heat treatment.

R. C. Hurst (Euratom, Petten, Netherlands): My question concerns your experiences with CVD coatings because apart from economic problems these coatings appear to be very promising.

Reply (Schweitzer, MTU-München, for Dr Betz): The chemical vapour deposition (CVD) coating processes must be divided into two types, those resulting in overlay coatings and those forming diffusion-type coatings.

The CVD diffusion-type coatings are very similar to those formed by a pack process. As the pack process is less expensive, few real CVD coatings are used, e.g. CVD silicide coatings for turbine blades of stationary engines and CVD chromized coatings for compressor blades, but there are some more CVD pack processes.

On the other hand, the CVD process is very suitable for depositing coatings in cooling holes and there is an increasing demand for its use. CVD overlay coatings, where all the elements in the coating are deposited from the vapour phase, e.g. titanium or molybdenum silicides, are currently in the same stage of development and application as coatings for refractory metals.

For wear resistance, CVD overlay coatings are already finding considerable application, e.g. carbides, oxides, nitrides and borides.

[See also Discussion of P. Felix's paper (Chapter 13)—ED.]

13

Coating Requirements for Industrial Gas Turbines

P. C. FELIX

Brown Boveri & Cie, Baden, Switzerland

ABSTRACT

This paper summarizes the main requirements which have to be fulfilled by protective coatings to resist corrosion of the blades of industrial gas turbines. As the most important conclusion it was found that a coating should never be developed or evaluated alone but always as one part of the coating–base alloy–blade system.

INTRODUCTION

Protective coatings have been used in industrial gas turbines, at least by some manufacturers, for more than 20 years. This could mean that the development of such coatings has already been completed a long time ago. The fact, however, that during this meeting 12 presentations dealing with corrosion-resistant coatings will be given, shows that a lot of work has still to be done in this area. This is mainly caused by the continuous change in the requirements which have to be met by these coatings—first of all by the increasing temperatures in today's and future turbines. The most important requirements for corrosion-resistant coatings for industrial gas turbines are as follows:

corrosion resistance,
erosion/impact resistance,
thermal stability,
mechanical strength,
adhesion,
influence on base alloy properties,
influence on blade functions,
economics.

P. C. Felix

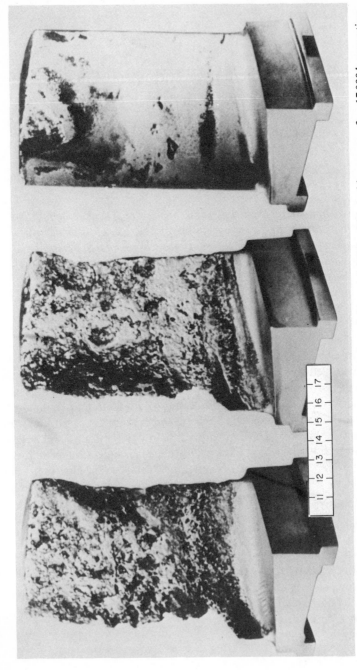

Fig. 1. Uncoated, aluminium and chromium coated (from left to right) Nim 80A turbine vanes after 17 000 h operation

This list shows that it is important not to select or evaluate coatings alone but always as part of the coating–base alloy–blade system. This statement seems to be trivial; however, there are still coating developments looking only for the highest possible corrosion resistance without evaluating any other of the above requirements.

CORROSION RESISTANCE

In stationary gas turbines burning primarily fossil fuels such as natural gas distillates, crude oil, etc., sulphidation initiated by sulphate deposits is the prevailing corrosion mechanism. This important fact has to be considered in the evaluation. Aluminium base coatings applied by different methods have been used in jet turbines for some time but they do not give the required long-term protection under sulphidation conditions. Sodium sulphate reacts with the aluminium oxide, which is the actual protective layer of Al coating, forming non-protective sodium aluminate (1). Short-term protection might be possible with some Al-base coatings if the attack is not too severe. In other cases, however, Al-base coatings may behave like uncoated blades and give no protection at all, as is shown in Fig. 1. The illustration also shows a chromium coating exposed to the same corrosive conditions revealing almost no attack.

Protective coatings are used for blades of industrial gas turbines in order to extend their corrosion life. This is done by reducing the corrosion rate at the blade surface, as is shown schematically in Fig. 2. It must be pointed out

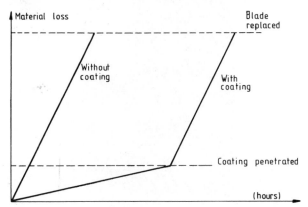

FIG. 2. Schematic comparison of corrosion life of a coated and an uncoated turbine blade (IN-713)

that the coating alone is not responsible for the corrosion life of the blade; the coating only delays the initiation of corrosion of the blade material. Overall, this combination results in a longer corrosion life of the blade. Therefore it is important to select blade alloys with the highest possible corrosion resistance. It would be impossible to use refractory metals such as molybdenum as blade material for industrial gas turbines with expected blade lives of 30–50 000 h. Besides, it has to be mentioned that up to now no commercial coating is available for which the absence of pores or other coating defects can be guaranteed beyond any doubt.

EROSION RESISTANCE

It is impossible to prevent particles of different sizes hitting the coated blade surface. The coating should not be completely destroyed by this process,

FIG. 3. Simulation of a coating defect, showing corrosive attack underneath the coating

leaving the base alloy unprotected from the corrosive attack. Ductile coatings are advantageous in this respect because an impact would deform but not destroy them. Figure 3 shows the effect of a local coating defect. Starting from the hole which was drilled through the coating to simulate the impact of a hard particle, the corrosion attack penetrates under the coating, leaving the coating unattacked but still destroying the blade underneath it. Pores, which are possibly already present in the virgin coating, may have the same effect as an impact. It is therefore important that the coating has no defects or open pores penetrating from the surface to the unprotected blade alloy.

THERMAL STABILITY

Coatings should be stable under any service condition, so that the protective elements of the coating responsible for the corrosion resistance remain in the coating and do not get 'lost' by diffusion into the base alloy. Such diffusion may also have a detrimental effect on the mechanical properties of the blade alloy, as will be shown later. The importance of evaluating alloy and coating as a system may be seen from the following field case. Figure 4 shows a first-stage turbine vane in IN-713 LC coated with a NiCrSi plasma coating (2) after 5000 h operation. The blister visible at the leading edge is the starting point of a corrosion attack. The metallurgical investigation showed that aluminium, of which 6 % is present in the base alloy but not in the coating, had diffused into the coating until a concentration had been reached high enough to destroy its proven sulphidation resistance. IN-738 LC, however, with only 3.4% Al was protected successfully, as can be seen from Fig. 5, and has been in operation for over 20 000 h without any visible corrosion.

MECHANICAL STRENGTH

The coating must be able to withstand all the stresses which can occur on the blade surface during operation of a turbine. If cracks in the coating were caused by creep stresses it could even be that coated blades would fail earlier than uncoated blades because these cracks could be starting points of a local attack, as was shown in Fig. 3. Cracks in the coating caused by fatigue stresses can also be dangerous because they may initiate fatigue fractures of the blade. In this respect it is important to have coatings and base alloys

P. C. Felix

FIG. 4. Corrosion initiation in an IN-713 LC turbine vane with a NiCrSi plasma
coating after 5000 h operation

FIG. 5. IN-738 LC vane with NiCrSi plasma coating after 20 000 h operation

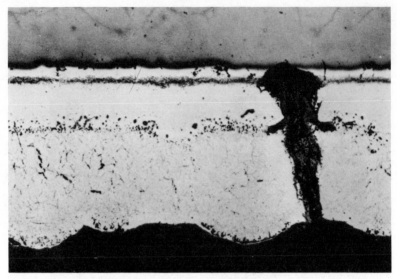

FIG. 6.   Fatigue crack in protective coating causing local corrosive attack

with similar thermal expansion coefficients in order to minimize the risk of thermal fatigue damage in the coating during start-up and shut-down procedures. Also, these cracks can be either serious when they proceed into the blade or just undesirable when they stop in the coating blade boundary layer and lead to local corrosion (see Fig. 6), but not to a thermal fatigue blade failure.

ADHESION

The coating must be locked to the substrate material so that neither mechanical nor thermal stresses cause it to flake off. General experience has shown that usually only coatings which are bonded to the base alloy by diffusion can meet this requirement. For multilayer coatings it is obvious that the same requirement must also apply to each single layer.

INFLUENCE ON BASE ALLOY PROPERTIES

The importance of the effects of the coatings on the mechanical properties of the blade alloys has only recently been recognized, at least for stationary gas turbines. It seems that due to the increased temperatures the difference between the necessary strength and the strength available has become smaller and smaller, i.e. in today's turbines the material properties are better utilized than in older turbines. On the other hand it seems that the modern superalloys, especially developed for the highest possible strength, are more sensitive to composition, heat treatment, etc., than older materials.

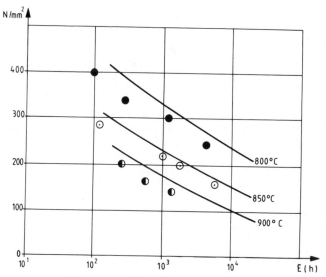

FIG. 7.   Creep rupture test of CVD Si-coated IN-738 LC showing considerable losses in strength. Curves = average data for uncoated IN-738 LC

As possible harmful effects of coatings, the diffusion of elements from the coating into the base alloy and the coating cracks as initiators of blade failure, have already been mentioned. Of special interest also is the effect of the heat treatment which is part of the coating application. IN-738 LC, one of today's standard alloys for industrial gas turbines, was coated (a) with a chemical vapour deposition (CVD) silicon coating (3) and (b) with a plasma-sprayed (2) NiCrSi coating. Both coatings need a heat treatment for proper bonding of coating and substrate. As can be seen in Fig. 7 it was found that the CVD silicon process reduced the creep rupture life of IN-738 LC by up to 60% of its original value. Since uncoated specimens subjected to the same heat treatment as that in the coating process showed the same effect, it was concluded that the heat treatment and not the Si coating was

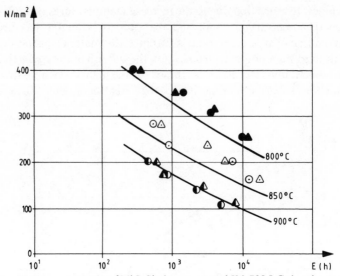

Fig. 8. Creep rupture test of NiCrSi plasma coated IN-738 LC showing no losses in creep strength. Triangles = coated and heat treatment, circles = heat treatment only

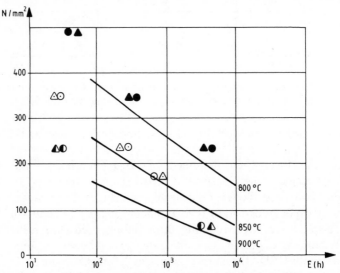

Fig. 9. Creep rupture test of CVD Si-coated Nimonic 105 showing no losses in creep strength. Triangles = coated and heat treatment, circles = heat treatment only

responsible for the problem. The plasma-sprayed specimen, on the other hand, did not show any reduction in strength, as can be seen in Fig. 8. By coating Nimonic 105, an older gas turbine alloy, with the above Si coating (see Fig. 9) no reduction in creep strength was found. It was therefore concluded that the heat treatment cycle did not generally produce adverse effects, but only in the case of certain alloys. We conclude from these tests that similar investigations, not only for creep rupture but also for low cycle fatigue (LCF) and high cycle fatigue (HCF), etc., have to be conducted for any potential coating–base alloy combination.

INFLUENCE ON BLADE FUNCTIONS

Not as critical as the effect mentioned above, but still to be considered, is the effect of the coating on the functions of the blade. It seems to be obvious that the coating should not change the shape of the blade aerofoil in such a way that the aerodynamic behaviour of the blade should be disturbed. It should also be mentioned that a high surface quality has to be maintained because rough surfaces can considerably affect the fluid dynamics and therefore the performance of the gas turbine. In general, it can be said that a coating should have a surface quality as good or better than a precision cast blade.

ECONOMICS

Last but not least, economic factors have to be investigated when evaluating protective coatings. As shown in the discussion on corrosion resistance, coatings are usually not an absolute necessity for running a gas turbine but a measure to extend the blade life. Therefore it has to be calculated whether the savings due to a lower blade consumption, less maintenance costs, etc., justify the coating or not. Statements made by potential coating suppliers that the coating will allow the use of dirty and less expensive fuel, have to be investigated carefully because they may be misleading and in some cases even wrong. As was shown above, dirtier fuels and the associated more corrosive conditions will also influence the corrosion life of the coating as well as that of the blade alloy.

In conclusion, it may be a true statement to claim that a coating doubles the life of a blade but it is wrong to claim that a coating increases the blade life by a certain number of hours. Therefore, in every case it has to be

investigated whether a lower grade fuel can be burned just because of th
use of coated blades or whether other precautions would need to be taken i
such fuels were used.

SUMMARY

The most important requirements which have to be met by coatings fo
industrial gas turbines have been discussed. As the most importan
conclusion it was found that a coating should never be developed o
evaluated alone but always as a part of the coating–base material–blad
system. To minimize the risk of unexpected results and wasted mone
during the development and evaluation of new coatings, test programme
should be established to cover all of these requirements.

For the development one should be aware of the fact that in most cases i
is not possible to completely fulfil all these requirements. Instead a coatin
should be developed that fulfils the combination of the requirements in a
optimum way.

REFERENCES

1. E. Erdoes, M. Semlitsch and P. C. Felix (1972). *J. Mat. Tech.*, **3**(4), 193–7.
2. M. Villat and P. C. Felix (1976). *Sulzer Tech. Rev.*, **58**(3), 97–104.
3. P. C. Felix and H. Beutler (1972). Proc. 3rd Conf. on Chemical Vapo
   Deposition (Salt Lake City), pp. 600–17.

DISCUSSION

J. A. Klostermann (Eindhoven University of Technology, Th
Netherlands): Dr Felix showed a micrograph of a plasma-sprayed laye
which had almost no porosity. I would like to ask him: (1) what is hi
interpretation of the dark spots on the micrograph, and (2) is he acquainte
with the technique developed by Houben and Zaat? This technique make
it possible to get clean layers with very low porosity and good adhesion b
plasma spraying without further treatment.

Reply: (1) The dark spots in the micrographs of plasma coatings ar
carbides. (2) Yes, our plasma-spraying people are aware of the techniqu
developed by Houben and Zaat. However, we do not think that plasma

coatings for turbine blades could be used without any further treatment, for the following reason. As mentioned in my presentation, we have found that blade coatings in general need a diffusion bonding to the substrate material in order to meet the adherence requirement. The heat treatment which we apply to our plasma coating is therefore needed for getting this diffusion bonding as well as for closing all the pores in the sprayed layer.

R. C. Hurst (Euratom, Petten, The Netherlands): Is it possible that corrosion of coatings in stationary turbine applications can trigger a hot corrosion reaction in a susceptible alloy? I have certainly seen in rig tests that aluminide-coated Nimonic 105 behaves in an inferior manner to the uncoated material. However, the intermetallic layer in this coating is, of course, not resistant to sulphur-bearing environments.

Reply: We have not seen yet such acceleration effects with the type of coating that we are using in our turbines. We have tested aluminide coatings in engines and have found in the worst cases that the coated blades behaved as badly but not worse than uncoated blades. In some cases coated blades looked worse than uncoated because a coating defect or a foreign object impact, etc., had destroyed the coating locally. This initiated local attack on the base alloy which was sometimes more dangerous than the uniform attack on an uncoated blade.

R. C. Hurst: My next question concerns your experiences with CVD coatings because, apart from economic problems, these coatings appear to be very promising.

Reply: I agree that CVD coatings in general have some interesting properties such as uniformity, high density, etc. For the CVD/silicon-base coatings, developed especially as protective coatings for turbine blades, we found high corrosion resistance and almost all the desirable properties listed in my presentation. More details are to be found in two earlier publications (refs. 1 and 2 of this paper). However, we did run into some problems, too, and these were basically responsible for us not putting these coatings into full-scale production. First, it was found that these silicides were quite brittle and sensitive to impact. A number of blades with Si coatings were removed from turbines with no attack at all in one area and heavy corrosion in another area where impact damage had occurred.

Another more general problem was the masking of the root and internal cooling passages. We did not find a masking method which was effective against gases at temperatures up to about 1000 °C to prevent coating of the root; this was necessary for design reasons.

[See also Discussion of W. Betz's paper (Chapter 12)—ED.]

A. Rahmel (DECHEMA, Frankfurt/Main, Germany): I wondered that your catalogue of requirements did not contain the reparability. It may be that this requirement is included in one of the other points.

Reply: I would call the reparability of a coating a desire but not a requirement. The reconditioning of blades, including recoating, which is common practice in aero turbines is a new business for stationary turbines. In such cases, however, we do not repair the coating but strip and recoat the blade completely. Another aspect of coating repair are the small defects such as pores, cracks, etc., which are found on new blades. Whether such faults can be repaired or not is basically a question of the application of the coating. Plasma-sprayed coatings, for example, are suitable for local repair work, but CVD coatings, for example, are better stripped completely and then reapplied. The ideal coating, however, would not need any repair because it would be self-healing, which is another desire but not really a requirement.

14

Phase Stability of High Temperature Coatings on NiCr-base Alloys

U. W. Hildebrandt, G. Wahl and A. R. Nicoll

Brown Boveri & Cie AG, Heidelberg, West Germany

ABSTRACT

All the requirements for good high temperature behaviour are met by coatings which are based on finely dispersed particles in a high melting point ductile matrix, provided that the particles contain corrosion-inhibiting elements such as Si or Cr.

The continuous release of one or more of these elements is of primary importance. Unavoidable ripening effects are influenced strongly by the elements which diffuse into the coating from the substrate.

The equilibrium structure may be influenced by variation of the coating thickness, by the use of interlayers and by changes in the matrix composition.

INTRODUCTION

This chapter investigates the long-term structural stability of two silicon-containing coatings on Ni, Co and NiCr-base alloys.

The coatings are all characterized by a dispersed phase which is either Si or Cr rich and which is embedded in a ductile homogeneous matrix. Such coatings have been shown to be very corrosion resistant, but there is a danger that structural changes due to ripening can take place. In the worst case these can lead to the formation of a duplex coating from which little mechanical stability can be expected.

EXPERIMENTAL

The coating with the Si-rich phase (hereafter referred to as the silicide coating) was produced using chemical vapour deposition and is intended to find application up to temperatures of 1000 °C (1). The coating with the Cr-rich particles, in the form of chromium boride, referred to as the boride coating, is a plasma spray coating (2) with a maximum working temperature of 900 °C.

Ni and Co were used at first as substrates followed by Inconel 738 LC (trademark, International Nickel), a common Ni-base superalloy, and a eutectic directionally solidified Co-base alloy 'C73' (3, 4). For reasons of mechanical compatibility, in the case of the silicide coating, both superalloys were sprayed with a Ni coating before siliconizing took place.

Rectangular-shaped specimens of the coated substrate materials were furnace-annealed in air for up to 1200 h.

During the metallographic investigation, special attention was paid to the stability of the structure and to the changes in elemental concentration.

Elemental concentration profiles were determined by point measurements in a microprobe analyzer and used to identify the phases present.

In the case of the boride coatings, quantitative size and distribution analyses of the CrB particles were carried out using a point-counting technique. The volume fraction, mean diameter and mean spacing of the particles as a function of the exposure time at temperature were determined.

RESULTS

Si Coating—Silicide-coated Nickel

With respect to the NiSi binary system shown in Fig. 1, a eutectic exists on the Ni-rich side which consists of a Ni/5 wt % Si solid solution and the nickel silicides Ni_3Si and Ni_5Si_2, respectively. Figure 2 shows the lamellar eutectic with the light-coloured silicides and the darkly etched γ-solid solution. The base material, pure Ni, can be seen to the right of the silicides. During exposure at temperature, the silicon diffuses from the eutectic structure into the γ-Ni, as shown by the concentration profiles in Fig. 2.

An interesting effect is the non-stoichiometric composition of the silicides, which only reaches a maximum level of 13.5 % Si on the outer surface which consists of Ni_5Si_2 (Fig. 3).

With increasing exposure time, separation of the eutectic structure occurs, this being almost complete after 200 h. Ripening effects occur such that the original eutectic grains transform to Ni_3Si, while on the surface a relatively thin layer of the very brittle Si-rich Ni_5Si_2 phase is formed, which can easily break away.

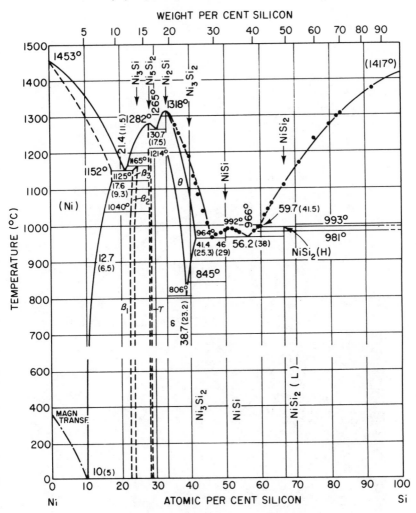

FIG. 1.   The NiSi binary phase diagram (from M. Hansen, ref. 5)

Silicide-coated Co

Figure 4 shows the CoSi binary system. Analogous to the case of NiSi, as shown above, a eutectic exists on the Co-rich side between the Co/7 wt % Si phase and Co_3Si. The Si richer Co_2Si is to be expected in the same way. Figure 5 shows the base material after coating and various heat treatments;

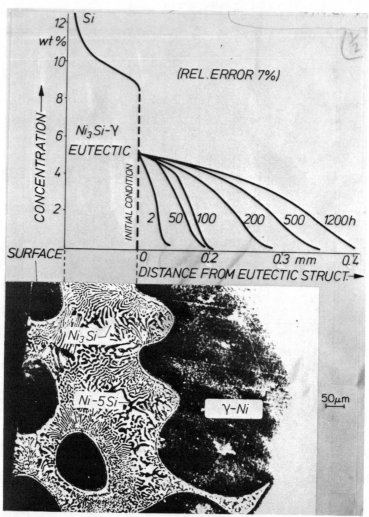

FIG. 2. Initial condition and Si-profiles for various exposure times for siliconized Ni

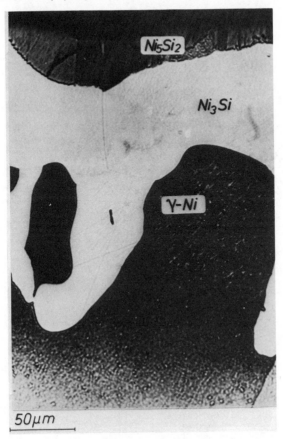

FIG. 3. Structure of siliconized Ni after 200 h/1000 °C in air

the Co/7 wt % Si solid solution is etched light in colour and the silicides dark. As for the coatings on Ni, concentration profiles in the base material for various exposure times are shown. A Si-rich Co_2Si silicide occurs at the eutectic boundary. Figure 6 exhibits structural ripening after long-term exposure but this is less marked than with the Ni silicides. The eutectic structure remains intact.

Ni Silicides on IN-738 LC

As has been mentioned already, the IN-738 LC specimens were first plasma sprayed with a Ni layer of approximately 100 μm thickness; the specimens were then Si coated in the normal manner.

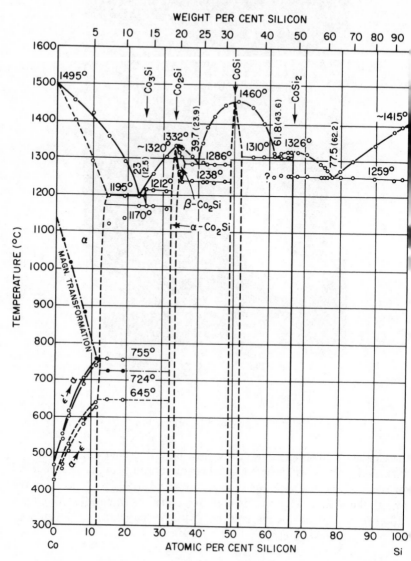

FIG. 4.   The CoSi binary phase diagram (from M. Hansen, ref. 5)

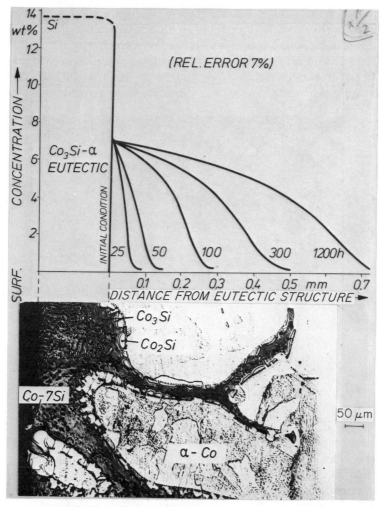

FIG. 5.   Initial structure and Si profiles after various exposure times for siliconized cobalt

Figure 7 shows the initial coated condition with the element concentration profiles. The coating exhibits the structure and elemental profiles of the model substance Ni with Si. However, larger grains of saturated solid solution occur in this case. As the profiles show, the γ-Ni zone is slightly enriched with Co, Cr and Al during siliconizing. It appears that Kirkendall holes have been formed which are visible along with the pores which form during the coating process. Figure 8 shows the coating after an exposure of 300 h at 1000 °C. A clear separation of the eutectic structure occurs which is always associated with breakaway of the coating.

The following layers occur from the outside towards the specimen centre: (a) a brittle Si-rich silicide Ni_5Si_2, (b) Ni_3Si, and (c) γ-Ni with a very heterogeneous structure, voids and inclusions being formed.

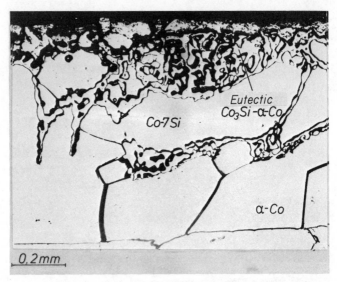

FIG. 6.   Structure of siliconized cobalt after 1200 h/1000 °C in air

The silicide coatings break down very easily during further exposure, particularly because of the thermal stresses which arise during cooling. This leaves a coating of γ-Ni solid solution (Fig. 9) with oxidized grain boundaries. The elemental concentration profiles exhibit the following characteristic changes compared with the initial condition: (a) a uniformly higher Si content of 7 wt %, which corresponds approximately to the maximum solubility in Ni, and (b) Cr and Co maxima in the base material and a sharp drop in Si content at the coating boundary.

The latter point indicates that Cr and Co are hindered from diffusing outwards by the Si which diffuses into the IN-738 LC. This effect is advantageous for two reasons. First, the diffusion of Si from the coating means less Si in the coating and less embrittlement of the base material, and secondly the Cr peak which forms through the uphill diffusion represents a Cr reservoir which is available for the further development of a Cr_2O_3 scale.

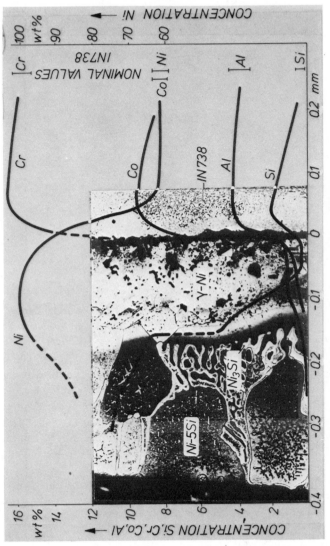

Fig. 7.   Structure and elemental profiles for IN-738 (NiSi), initial condition

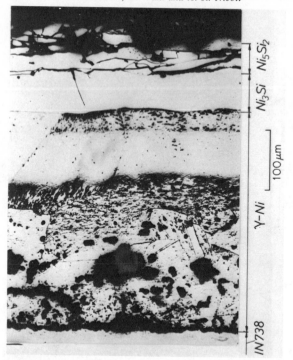

FIG. 8.   Structure of IN-738 (NiSi) after 300 h/1000 °C in air

Silicide-coated Ni on C73

A further substrate material that was investigated was the directionally solidified eutectic superalloy, C73, which consists of Cr_7C_3 fibres in a CoCr matrix. This alloy exhibits excellent high temperature mechanical properties and corrosion resistance. The specimens were plasma sprayed with Ni and coated with silicon as mentioned above.

In order to study the influence of cobalt on the structure, Ni coatings of $\sim 600\,\mu m$ and $\sim 300\,\mu m$ were used. As the cobalt can enter the coating solely by diffusion, the thinner nickel coating simulates, after only a short exposure time, the effects obtained with thicker coatings after long-term exposures, due to its relatively high Co content. Thus, it shows the structure expected after exposure times of ~ 5000–$10\,000$ h, experienced under practical conditions in gas turbines.

The initial structure of the 0.6 mm thick Ni layer on C73 corresponds to that of the model system NiSi. Figure 10 shows the structure after 1000 h. Ripening processes occur and, in particular, a continuous Ni_3Si layer

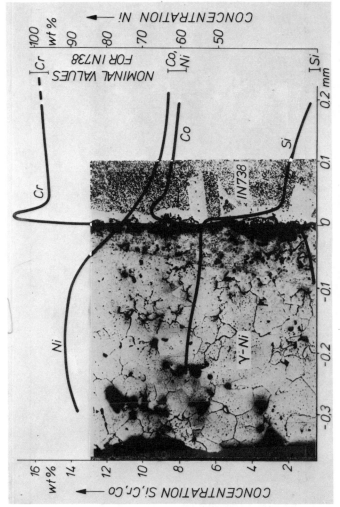

Fig. 9. Structure and elemental profiles in IN-738 (Ni/Si) after 1000 h/1000 °C in air

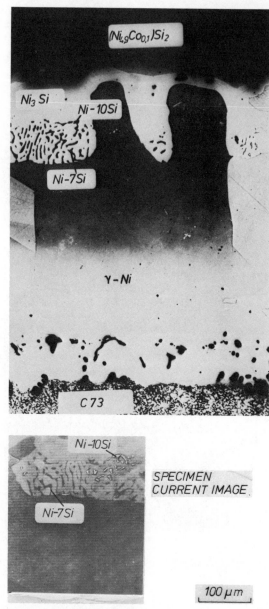

Fig. 10.   Structure of C73/NiSi after 1000 h/1000 °C in air (600 μm thick Ni layer)

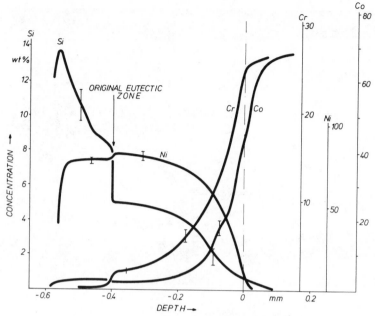

FIG. 11. Elemental concentration for C73/NiSi after 1000 h/1000 °C (600 μm thick Ni layer)

forms. A characteristic feature is the inclusion of Co in the surface layer with silicides of the type Ni_5Si and the occurrence of two supersaturated phases of differing silicon contents clearly defined by the microprobe analysis results. The adhesive strength is reduced by the occurrence of a zone of pores which develops in the boundary layer.

Figure 11 shows the elemental concentration. In contrast to the behaviour of IN-738/Si described above, the diffusion of Cr or Co is not hindered here. The comparison shows that Si contents $> 4 \%$ are apparently necessary. In the porous zone, Cr and Co contents of approximately 10% are found. The same distribution occurs with the thinner Ni layers, the structure being determined essentially by the diffusion of cobalt into the coating layer, as shown in Figs. 12 and 13. A continuous layer of lamellar carbides of the type $Cr_{23}C_6$, with large amounts of Ni, Co and Si, is formed in the coating, from the carbon in the base material carbides. Ni_3Si_2 silicides do not occur at all in this case. On the other hand, very Co-rich (up to 20 at %) Ni_5Si_2 silicides are formed. It is believed that the presence of these would determine the corrosion resistance after very long exposures.

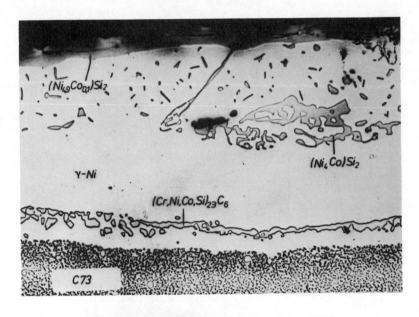

FIG. 12.   Structure of C73/NiSi after 1000 h/1000 °C in air (300 μm thick Ni layer)

Figure 13 shows the concentration profiles of Co and Si in the initial condition and after 1000 h exposure. The Co maximum is found in the carbide zone, as shown in Fig. 12. As the Si content of the matrix is again higher than 4 %, the diffusion barrier mechanism described above would appear to be the reason for the existence of the carbide zone. In general, the Co content in the coating amounts to more than 20 %.

Figure 14 exhibits the corresponding concentration profiles for Ni and Cr. The amount of Ni in the coating is reduced from 90 to 50 % and again a maximum in the Cr concentration is observed. This provides a Cr reservoir for the formation of chromic oxide scale.

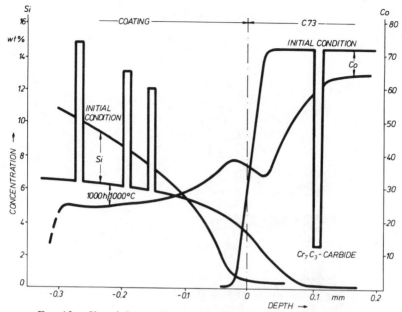

FIG. 13. Si and Co profiles on C73/NiSi (300 μm thick coating)

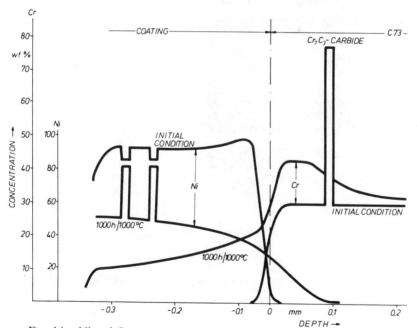

FIG. 14. Ni and Cr concentrations on C73/NiSi (300 μm thick coating)

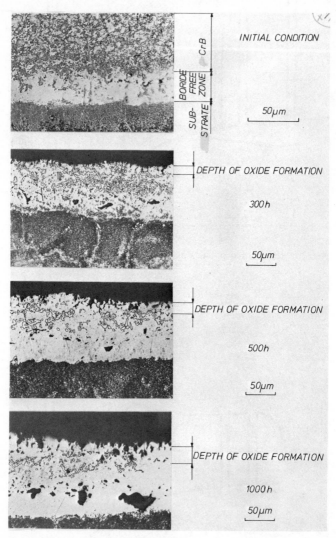

FIG. 15.   Boride coating structural changes after exposure in air at 900 °C

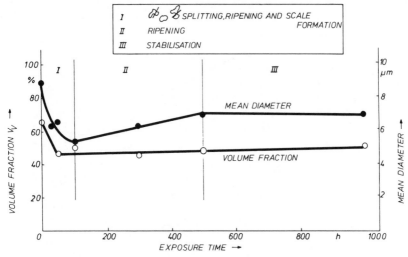

FIG. 16. Volume fraction and mean diameter of the CrB particles as a function of the exposure time at 900 °C in air

NiCrSiB Coating on IN-738 LC

As mentioned previously, the boride coating comprises a homogeneous NiCrSi matrix with a defined amount of dispersed chromium boride. The coating exhibits very good corrosion resistance. It will be shown in a semi-quantitative manner that with this coating a Cr-rich phase can supply Cr continuously over a long period of time in accordance with the Cr reservoir hypothesis.

Figure 15 shows the initial condition and the structures formed after exposures of up to 1000 h at 900 °C. It can be clearly seen that, as the thickness of the oxide layer increases, that of the chromium boride zone decreases.

The results of a quantitative metallographic analysis are shown in Fig. 16. The volume fraction and mean diameter of the borides are given for various exposure times. The three regions of the curves can be described as follows.

Region I up to ~100 h is characterized by a reduction of both of these parameters. The reduction of the mean diameter is explained by the large particles splitting into smaller ones. The volume fraction reduction from 65 to 45 % is determined by the oxide scale formation. In region II (up to ~500 h), Ostwald ripening occurs (until a stable condition, region III, is reached above 500 h).

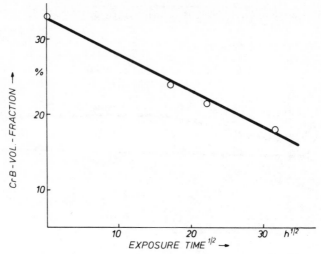

FIG. 17.   CrB particle fraction in the coating as a function of the exposure time in air

A plot of the volume fraction (which becomes smaller with the reduction in width of the boride zone) against the square root of the exposure time, gives a straight line (Fig. 17). Combined with the parabolic oxidation law, this indicates quite strongly that the Cr-rich particles guarantee a uniform oxide scale and thus produce the basic requirements for the excellent high temperature corrosion resistance.

REFERENCES

1. G. Wahl. (1977). 'Siliconizing of superalloys', Brown Boveri Co. Report, Mannheim.
2. H. Haas (1977). *Plasma-sprayed coatings, Brown Boveri Co. News.*, Nachrichten, Mannheim, in press.
3. E. R. Thompson and F. D. Lemkey (1973). United Aircraft Report No. M110544-2.
4. P. R. Sahm (1975). *High Temperatures–High Pressures*, **7**, 241.
5. M. Hansen (1958). *Constitution of Binary Alloys*, McGraw-Hill, New York.

DISCUSSION

W. Möller (Babcock–Brown Boveri Reaktor GmbH, Mannheim, Germany): In your figures you indicated a Si concentration of 8 %. Is this concentration sufficient for corrosion protection? I remember service tests in the early sixties of blast furnace gas turbines with Si-coated blades. The Si concentrations were about the same. These tests were disappointing because the coatings failed, as did the aluminized coatings which were also tried.

Reply: The properties of the silicide layers depend on the base material on which they are formed for two reasons:

(1) The silicide layers are mainly produced by interaction of the silicon with the base alloy. The Si content and the structure is therefore determined by the deposition process *and* the base material.

(2) The coating/base material system is a simple case of a composite material (see Workshop, 'Verbundwerkstoffe', *Z. Werkstoff*, **9,** 76). The chemical and mechanical properties of the layer are influenced by the properties of the base material.

For these reasons it is difficult to transfer experience from one alloy coating system to another.

15

Ductile–Brittle Transition of High Temperature Coatings for Turbine Blades

A. R. NICOLL, G. WAHL and U. W. HILDEBRANDT

Brown Boveri & Cie AG, Heidelberg, West Germany

ABSTRACT

The optimization of the mechanical properties of high temperature coatings includes the determination of the ductile–brittle transition temperature (DBTT). In spite of the relatively small amount of information available from the literature, it can be shown that DBTT values can vary according to coating type between DBTT < room temperature and DBTT $\simeq$ 1000 °C. The purpose of this investigation is to understand the ductile–brittle behaviour of Si-base coatings in terms of the influence of a number of parameters. Special attention will be given to the mechanical compatibility between chosen base materials (IN-738 LC, Nimonic 105 or IN-939) and the coatings, in particular with respect to crack initiation and growth at temperatures up to 1000 °C.

INTRODUCTION

Gas turbine coatings are subject to various forms of mechanical stressing and corrosion; the latter is described in other chapters in this book and elsewhere in the literature. Mechanical stressing includes the following: (1) centrifugal stresses (tensile stresses $< \sigma_{0.2}$); (2) tensile and compressive stresses which are brought about by differing expansion coefficients; (3) impinging particles; and (4) high cycle and thermal fatigue.

For the coatings to remain undamaged, they have to satisfy the following requirements: (1) the coating should exhibit a certain tensile strength, and (2) the coating should exhibit enough ductility to resist very high momentary shock loads.

233

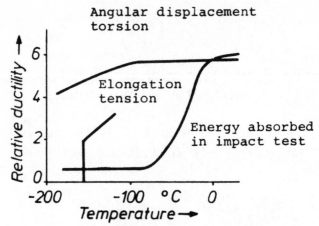

FIG. 1.   The ductile–brittle transition of iron as measured by three types of test: impact, tension and torsion (from Heindlhofer, ref. 2)

Consideration of the mechanical properties of a coating is complicated by its extreme anisotropy, and therefore it is not possible to consider the coating properties without including the coating base material.

The workshop 'Verbundwerkstoffe' (1) brought to light the fact that, in general, a coating has to be considered as a completely new material which is defined by its chemical composition, crystal structure and the special properties of the interface zone between the base material and the coating.

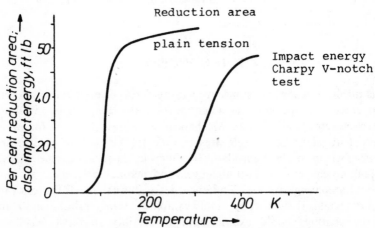

FIG. 2.   Ductile–brittle transition in a Bessemer rimming steel

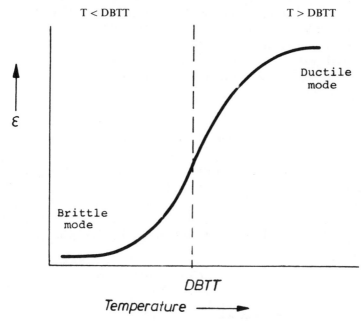

FIG. 3. Generalized curve for ductile–brittle transitions

The need for such investigations of the mechanical properties is shown by the frequent damage to turbine blading, where the coating either completely breaks away or exhibits cracking which leads to accelerated corrosion and blade failure.

The purpose of this investigation is to understand the ductile–brittle behaviour of a coating and to optimize it.

The general deformation behaviour of any material can be investigated by different methods, e.g. by impact, tension or torsion tests. Figure 1 shows a comparison of the results of all three tests for iron (2). The torsion test will not be dealt with here; in the first test the impact strength (or energy, J) is measured and in the second test the fracture behaviour is studied. The reduction of area (RA) and the elongation (L) are measured, as can be seen in Fig. 2 for a Bessemer rimming steel (3).

J, RA and L are temperature dependent for hexagonal and body-centred cubic metals and for many intermetallic compounds, and have the following characteristics.

The ductile and brittle modes are separated by a transition range—this is shown in Fig. 3. The curve is determined qualitatively by the values of J, RA and/or L in the brittle mode and in the ductile mode, as well as by the ductile–brittle transition temperature (DBTT). The behaviour shown in Fig. 3, with transition temperatures above room temperature, is found only for metals having hexagonal or body-centred crystal structures and not with FCC metals. These exhibit a ductile behaviour down to very low temperatures.

THEORETICAL CONSIDERATIONS

The microstructural features with which the ductile–brittle behaviour can be correlated qualitatively are shown in Fig. 4.

Mechanical loading causes the dislocations to move, to become pinned on all types of obstacles and defects and eventually to pile up. The increase in internal stresses can then either lead to cracking, or with increasing stress

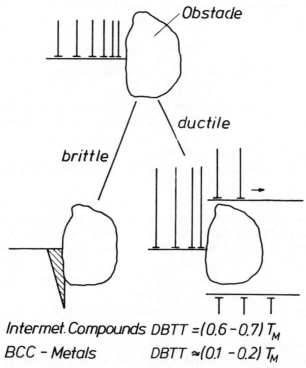

Intermet. Compounds DBTT =(0.6 – 0.7) T_M
BCC – Metals   DBTT ≈(0.1 – 0.2) T_M

FIG. 4. The influence of dislocation movement and mechanical loading

and temperature the activation energy for vacancy diffusion is such that the dislocations may climb on to other glide planes. This is possible, in particular, at the higher temperatures.

Ductile–brittle transition temperatures can be correlated empirically with the alloy melting points. For intermetallic compounds, Lowrie (4) found that all ductile–brittle transitions occur in the range

$$\text{DBTT} \sim 0.6 \ldots 0.7 T_m$$

where T_m is the melting point in degrees Kelvin. For BCC metals, ductile–brittle transition temperatures are lower:

$$\text{DBTT} \sim 0.1 \ldots 0.2 T_m$$

PARAMETERS WHICH CAN INFLUENCE DBTT

Ductile–brittle transition temperatures are dependent upon the various parameters discussed below. Chemical composition and microstructure have been deliberately excluded.

(1)   An important parameter is the strain rate $\dot{\varepsilon}$. DBTT decreases with increasing strain rate, as shown in Fig. 5 for tungsten (5). The empirical relationship $\text{DBTT} \propto \exp(-\dot{\varepsilon}^{0.09})$ fits the data well.

(2)   The transition temperature has been found to be strongly dependent upon surface condition, as shown in Fig. 6 for tungsten (5).

(3)   The transition temperature is dependent upon grain size in a sensitive manner (5), as shown in Fig. 7. In the case of tungsten, a

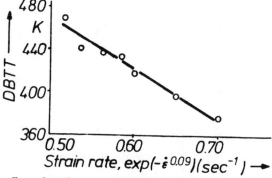

FIG. 5.   The effect of strain rate on DBTT of tungsten (from Stephens, ref. 5)

A. R. Nicoll, G. Wahl and U. W. Hildebrandt

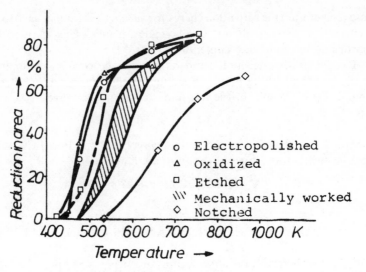

FIG. 6. The effect of surface condition on DBTT of tungsten (from Stephens, ref. 5)

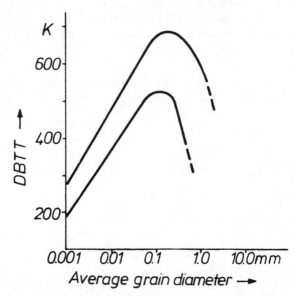

FIG. 7. The effect of grain size on the DBTT of tungsten

maximum transition temperature is observed with a grain size of 0.1 mm.

(4) The oldest known factor which can influence the DBTT is embrittlement by interstitial impurities (see para. 7 below), e.g. nitrogen embrittlement in chromium, hydrogen embrittlement in steels and copper, and oxygen and carbon embrittlement in tungsten. Figure 8 shows the latter effect (5).

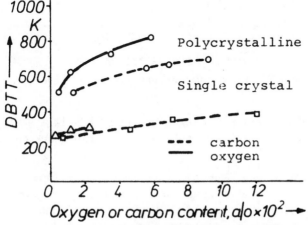

FIG. 8. The effects of oxygen and carbon on DBTT of single crystal and polycrystalline tungsten (from Stephens, ref. 5)

(5) Small particle additions can lower the DBTT, as shown in Fig. 9 for Cr–3 % ThO_2 (wt %) (6). In this case, the DBTT is lowered from 240 °C to 50 °C.

(6) Alloying elements have been found to change the ductile–brittle behaviour. An example of this effect is the ductility enhancement (solute softening) of Cr with rhenium (10), shown in Fig. 10. Other metals also show this effect in Cr, for example, Ru, Co and Fe.

(7) Isothermal exposure in air has been shown to increase the DBTT. Figure 11 shows the effect of exposure for 100 h at 1420 K for various coated specimens, compared with the uncoated substrate (8).

(8) A further important factor is the mechanical handling of the material, which can also cause the DBTT to vary. This point will be dealt with elsewhere.

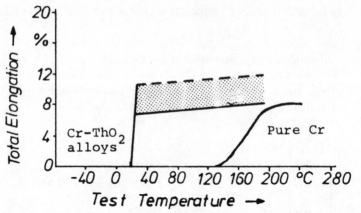

FIG. 9.   Tensile DBTT of as-rolled chromium and Cr–ThO$_2$ alloys (from Wilcox, ref. 6)

Although it is not intended that a model for DBTT be provided here, it should be mentioned that a viable model (9) must explain the following features:

(a)   the effect of alloying on the cohesive fracture stress,
(b)   the effect of alloying on solute softening (see item 6 above),
(c)   the effect of alloying on solid solution strengthening,
(d)   temperature and strain rate effects (see item 1 above) on the flow properties,
(e)   the effect of notches or cracks on ductility, and
(f)   the effect of strain hardening on ductility.

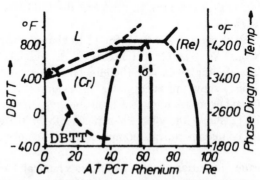

FIG. 10.   The ductile–brittle transition temperature and phase diagram for chromium–rhenium alloys (from Klopp, ref. 7)

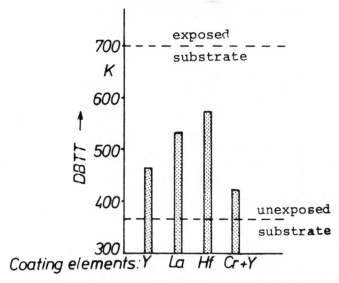

FIG. 11. The ductile–brittle transition temperatures of Cr/0.17Y uncoated; coated with Y, La, Hf and Cr + Y; and exposed isothermally in air for 100 h at 1420 K

DUCTILE–BRITTLE TRANSITION TEMPERATURES OF COATINGS

Compared with the uncoated substrates, the situation in this case is more complicated as the coating cannot be considered independently of the substrate. This means that in addition to the parameters which influence the coating, parameters must be included which relate to the base material and to the base material/coating interface.

A strain to fracture method (10) was chosen to measure the ductile–brittle transition, as cracking of the coating on the substrate is easily observable. This method of testing involves observation of the surface of a coated specimen during a tensile test with a telescope and calculation of the strain at which the coating breaks from the substrate or at which cracks appear. The experimental arrangement is shown schematically in Fig. 12. The various base material/coating combinations which were examined are given in Table 1. In general, this investigation was concerned with silicon coatings on nickel-base superalloys, coating being carried out either by chemical

242

A. R. Nicoll, G. Wahl and U. W. Hildebrandt

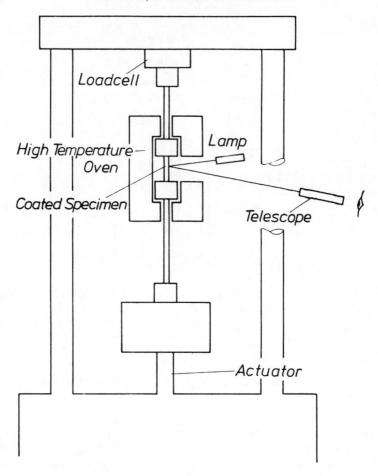

FIG. 12. Schematic of the experimental arrangement

vapour deposition (see Chapter 20 herein), glow discharge (see Chapter 20) or salt bath deposition (11).

In order to be able to compare results with those from coatings which are already in use, the coating DP-24 (Thyssen) on Nimonic 80A was tested. Other coating/base material combinations which did not exhibit a DBTT are given in Table 2. As has already been shown in Fig. 12, the testing was carried out using a servo-hydraulic testing machine at temperatures between ambient and 1000 °C with constant strain rate.

TABLE 1

Base materials, coatings and methods of coating used in this investigation

Base material	Coating	Method of preparation
IN-738 LC[a]	Silicide base	Chemical vapour deposition (hot wall)
Nim 105[a]	Silicide base	Chemical vapour deposition (hot wall)
IN-939[a]	Silicide base	Chemical vapour deposition (hot wall)
IN-738 LC[a]	Silicide base	Glow discharge
C73/2Al[b]	Silicide base	Salt bath
Nim 80A[a]	DEW-DP24[c]	Pack cementation

[a] Trademark, International Nickel.
[b] High temperature cobalt-base directionally solidified eutectic.
[c] Trademark, Thyssen.

RESULTS AND DISCUSSION

The curve for silicon-coated IN-738LC (CVD) is shown in Fig. 13. The test temperature is plotted against the strain to cracking (this being the plastic strain). In this particular case, three different zones can be identified. In the first zone, i.e. up to a temperature of approximately 500 °C, integrity is lost and the coating breaks away.

Similar cracking occurs up to a temperature of approximately 700 °C, but above 850 °C the coating/base material combination remains stable, without any surface changes, up to fracture of the base material. The high base material elongation of 18 % occurs as the normal heat treatment of 845 °C/24 h was not carried out. The specimens were examined metallographically. The coating composition will not be discussed here.

Figure 14 shows the first zone, which is characterized by loss of coating integrity. At room temperature, 300 and 450 °C, coating breakaway is observed in the elastic part of the stress/strain curve. Two effects can be observed—coating breakaway and microcracks in the coating.

Figure 15 shows the coating microstructures for the test temperatures of 600 and 700 °C in zone 2. The cracks seen in zone 1 increase in width and depth, movement being into the base material from the coating. No coating breakaway occurs in this zone, the combination remains intact. Zone 3 is characterized by pore formation, as seen in Fig. 16 for the two test temperatures of 850 and 1000 °C. Cracking in the coating is not evident. The base material, however, is cracked, and the cracks move into the coating.

The same method of testing was used for all of the above-mentioned

TABLE 2

Comparison of DBTT and brittle fracture elongation (ε_{min}) from the literature and this investigation

		Co-Al (35%)	Ni-Al (35%)	Silicon — Salt bath	Silicon — Glow discharge	Silicon — CVD	CoCrAlY	NiCrAlY	NiCoCrAlY	NiCrSi-base	DEW-DP24
Nim 80A	DBTT										450-600°C
	ε_{min}										2.8%
Nim 105	DBTT	Base material not known	Base material not known			300-600°C	Base material not known	Base material not known	Base material not known		
	ε_{min}					0.71%					
IN-713	DBTT	850-1000°C	700-800°C			450-700°C	22Cr/12Al DBTT: 650-850°C	DBTT: from 650°C	DBTT: from 480°C	<0°C	—
	ε_{min}	0.2%	0.4%			<0.2%	0.4%	0.8%	0.5%	—	—
IN-738	DBTT				600-850°C	450-600°C	15Cr/10Al DBTT: from 200°C			<0°C	—
	ε_{min}					<0.2%	0.4%				
IN-939	DBTT									<0°C	—
	ε_{min}										
C73	DBTT			600-850°C							
	ε_{min}			<0.2%							

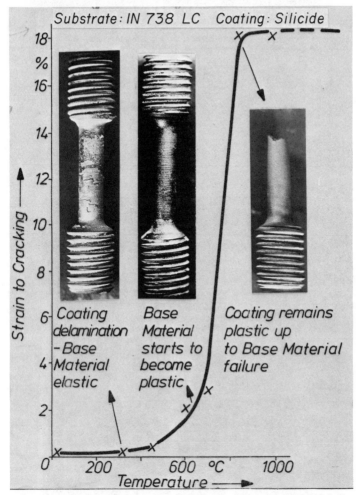

Fig. 13. The ductile–brittle transition for siliconized IN-738 LC

coating/base material combinations, the results being shown in Fig. 17. The strong influence of the base material on the ductile–brittle transition is made clear by this figure. Comparison of the results for the base materials Nim 105, IN-939 and IN-738 LC, all coated in the same manner with the same silicon-base coating, shows that the DBTT for Nim 105 lies approximately 200 °C lower than the DBTT for IN-939 and IN-738 LC, which lie almost together. It is also noticeable that base materials can raise

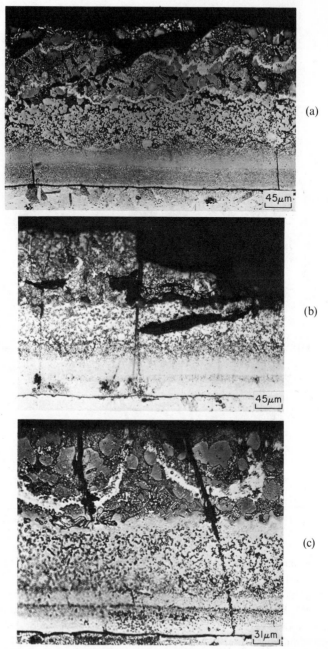

FIG. 14.   Microstructures of siliconized IN-738 LC in zone 1 ($T <$ DBTT)
(a) and (b) T = 300 °C (c) T = 450 °C; $\varepsilon = 0.34\%$

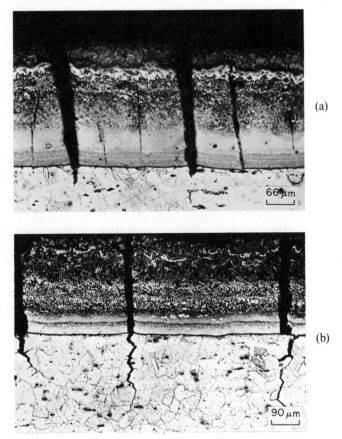

FIG. 15. Microstructures of siliconized IN-738 LC in zone 2 ($T \sim$ DBTT)
(a) T = 600 °C, ε = 2.01 %; (b) T = 700 °C, ε = 2.86 %

or lower the brittle mode elongation. This can be seen by the results for silicon coatings on Nim 105, which show approximately 0.7 % elongation at room temperature, compared with 0.2 % for similar coatings on IN-939 and IN-738 LC.

The combination Nimonic 105 and silicon-base coating has substantially superior properties to the other combinations, not only with regard to DBTT, but also with respect to the ductility in the brittle range (zone 1). However, the combination Nimonic 80A and DP24, which is used in practice, has a ductility higher by a factor of three. Curves for NiCrAlY and CoCrAlY coatings (12) on an unknown base material are shown for

248   *A. R. Nicoll, G. Wahl and U. W. Hildebrandt*

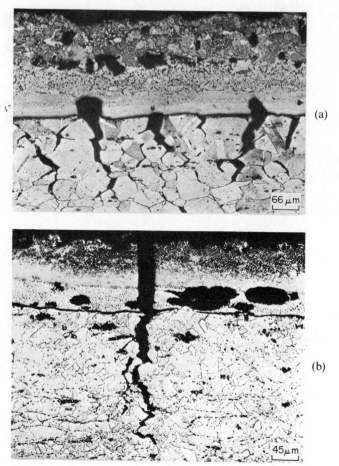

(a)

66 μm

(b)

45μm

FIG. 16. Microstructures of siliconized IN-738 LC in zone 3 ($T >$ DBTT)
(a) T = 850 °C, ε = 18 %; (b) T = 1000 °C, ε = 18 %

comparison. Lowering the aluminium concentration of CoCrAlY from
12 % to 10 % reduces the DBTT from 750 to 150 °C.

The varying amounts of strain required for cracking can be further
discussed by considering Fig. 18. This shows proposed stress/strain curves
for the coating and base material for the situations $T <$ DBTT and $T >$
DBTT. Also included are the expected load levels in turbines during the
start phase and during normal running. At $T <$ DBTT, the coating exhibits
brittle behaviour and coating breakaway occurs in the elastic range of the

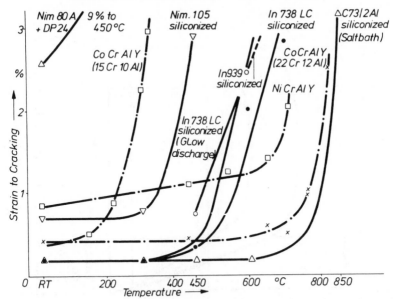

Fig. 17. The ductile–brittle transition temperatures of several protective coatings.
———·——— From Boone[12]

base material stress/strain curve. At $T >$ DBTT, the coating has become more ductile than the base material which fails at lower strains.

Table 2 shows transition temperatures from the literature and from this investigation for various coating/base material combinations. The DBTT and elongation in the brittle mode are given. Coatings which are already in use are shown on the right of the table. They do not exhibit (with the exception of Nim 105 + DP24) a transition above room temperature. The MCrAlY (the generic name—where M is Ni, Co, Fe or NiCo) coatings which are not yet in practical use in stationary gas turbines exhibit a rather high DBTT. However, the brittle mode elongation is sufficient (0.4–0.8 %).

CONCLUSION

The ductile–brittle transition temperatures shown in Table 2 can be compared with the empirical values given previously (DBTT $\sim$ $0.6\ldots0.7T_m$). The aluminide coatings with melting points of approximately $T_m \sim 1900$ K give a DBTT of

$$\text{DBTT (aluminide)} \sim 860\text{–}1050\,^\circ\text{C}$$

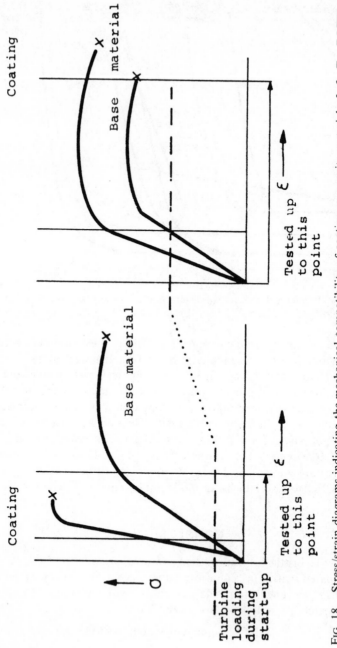

FIG. 18. Stress/strain diagrams indicating the mechanical compatibility of coating composite materials. Left, $T <$ DBTT; right, $T >$ DBTT

The melting points of silicides are lower, at approximately 1500 K. This gives

$$DBTT \text{(silicide)} \sim 600\text{–}800\,^\circ C$$

Both of these values are in good agreement with the maximum DBTT for these coatings. The DBTT of chromium coatings is lower in agreement with the relationship

$$\sim 0.1 \ldots 0.2 T_m$$

for BCC metals.

Of all the parameters we have considered which may influence the DBTT and the ductility, only one is of practical significance, namely, alloying with other components to reduce the activity of brittleness-inducing elements (silicon, aluminium). This has already been done with the CoCrAlY and NiCrAlY coatings. A further important parameter, which has not been dealt with here, is the coating structure, which is dependent on the method of coating. However, the influence of this parameter (coating structure) is limited in nature by the problem of structural changes and ripening, which is dealt with in the previous chapter.

ACKNOWLEDGEMENTS

The authors would like to thank Mr Sick for helping with the specimen preparation, Mr Beck for carrying out the tensile testing and Dr P. R. Sahm for helpful discussion. This work was supported in part by the German Ministry of Research and Technology (BMFT).

REFERENCES

1. Workshop 'Verbundwerkstoffe' (Konstanz, 1976), *Z. Werkstoff.*, **9.**
2. K. Heindlhofer (1934). Am. Inst. Min. Met. Engrs, Tech. Pub. No. 581, 7 pp.
3. A. H. Cottrell (1967). *An Introduction to Metallurgy*, Arnold, London.
4. R. Lowrie (1952). *Trans. Am. Inst. Min. Engrs*, (*J. Met.* **4,** 1093).
5. J. R. Stephens (1974). *Rev. Def. Behav. Mat.*, **1,** 31.
6. B. A. Wilcox *et al.* (1972). *Met. Trans.*, **3,** 273.
7. D. Klopp (1975). *J. Less Comm. Met.*, **42,** 261.
8. J. R. Stephens (1972). *Met. Trans.*, **3,** 2075.
9. W. W. Gerberich and Y. T. Chen (1976). 2nd Int. Conf. Mech. Behav., pp. 1191–95.

10. J. R. Ross and W. R. Cattin (1974). General Electric, Report 74CRD169 (August).
11. P. R. Sahm *et al*. Brown Boveri Research Lab., Heidelberg, unpublished.
12. D. H. Boone (1976). Airco Temescal.

DISCUSSION

M. Betz (MTU, München, Germany): The figure showing a coated specimen after tensile testing (Fig. 16, bottom) at 1000 °C reveals that the crack formation starts exclusively in the base material close to the coating. This would appear to indicate an influence of the coating on the base material.

Reply: It is unfortunate in this case that the figure does not show the whole cross-section of the specimen, as many of the cracks in the base material do not enter the coating. Consideration of the large pores in the coating directly adjacent to the base material would perhaps provide useful information on whether the coating influences the base material at such high temperatures. The figure showing a coated specimen after tensile testing at 850 °C (Fig. 16, top) indicates quite clearly that the cracks start in the base material/coating interface, but propagate to a greater extent in the base material, thus indicating that at this temperature the coating has the higher ductility.

16

Diffusion and Precipitation Phenomena in Aluminized and Chromium-aluminized Iron- and Nickel-base Alloys

E. FITZER and H.-J. MÄURER

Universität Karlsruhe, West Germany

ABSTRACT

The diffusion of aluminium into iron- and nickel-base alloys has been studied and the influence of the alloy composition on the structure of the diffusion layer and of the growth rates of the phases formed has been observed. A description is given of the formation of diffusion barrier layers on nickel-base alloys and of how their effectiveness depends on the concentration of active alloying elements in the base alloys. Possible methods of increasing the effectiveness of such layers are discussed.

INTRODUCTION

Aluminium diffusion-type coatings, the so-called aluminide coatings, are used to increase the high temperature corrosion resistance of nickel and cobalt superalloys in burner gases (for example on turbine blades (1)). In recent times they have also proved to be suitable as coatings on iron-base alloys in refinery desulphuration plants (2).

Figure 1 shows schematically how the corrosion resistance of coated alloys is improved by increasing the thickness of the coating. A disadvantageous side-effect is that the mechanical properties of the coated alloy can deteriorate (3). For example, it was shown that the long-term rupture strength and the thermal fatigue strength both decrease considerably with increasing coating thickness. One can obtain an increase in the average lifetime of the base alloy with the coating only if limited stresses are applied; in these cases the practical application of aluminide coatings is fully justified. In cases where very high mechanical loads are

253

applied, samples without coatings may exhibit a longer lifetime than coated materials.

Such a dependence of the specific resistance (meaning that resulting from the combination of the chemical and mechanical properties of the composite) on coating thickness is not only restricted to aluminide layers; it is a general observation when applying diffusion-type coatings made of chemically resistant, but brittle, elements.

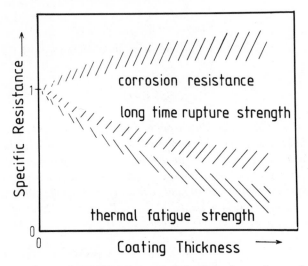

FIG. 1.   Dependence of specific resistance of aluminized samples on the coating thickness (from Betz *et al.*, ref. 3)

The impairment of mechanical properties, for example the decrease in strength of coated alloys, may be caused by several phenomena:

(1)   The properties of the base material may be altered by the heat treatment associated with the coating process.

(2)   The coating itself exhibits low strength but high brittleness.

(3)   The change of composition and preferred orientation of the base material, due to enrichment with the coating element, aluminium, by diffusion from the layer into the substrate. Embrittlement of the surface zones is the consequence of this enrichment.

In this chapter, item (3) the diffusion of aluminium into base materials of various compositions is studied. One can expect that, under comparable conditions, the amount of this inward diffusion of aluminium will depend primarily on the chemical composition of the base material and thus on the structure of the diffusion layer which is formed during preparation of the coating.

EXPERIMENTAL

For the demonstration of the influence of the nickel concentration of the substrate on the coating structure and on the phases formed, the alloys listed in Table 1 were used. These alloys were coated in a retort containing a

TABLE 1
Composition of iron-based alloys

Alloy	Composition (w/o)								
	Fe	Ni	Cr	Co	Al	Ti	Si	Mn	C
Remanit 4301	bal	10	20	—	—	—	—	—	0.07
Thermax 4841	bal	20	25	—	—	—	2.0	0.7	0.1
Thermax 4876	bal	32.5	21	—	0.4	0.4	0.5	1.2	0.08

pack mixture of $20w/o$ Al, $3w/o$ NH_4Cl and $77w/o$ Al_2O_3. The retort was not sealed and was heated in a tubular furnace continually flushed with hydrogen and open at one end. In this way it was possible to insert the samples to be coated immediately into the hot furnace, thus eliminating a long heating period. The samples were annealed between 950 and 1100 °C for up to 5 h and, after removal from the furnace, they were allowed to cool to room temperature in air. All the samples coated by the authors were made under these coating conditions (the coated nickel-base alloys shown in Fig. 2 were additionally heated after the coating process for 3 h at 1000 °C to reduce the hardness and brittleness of the aluminide layers).

The Ni–Cr model alloys, used to exhibit the influence of chromium in nickel-base alloys, were coated by Meinhardt (6) in a pack mixture containing $10w/o$ Al and $90w/o$ Al_2O_3. The samples were heated in a sealed retort in a hood furnace. The heating up time was taken into account when calculating the growth rate constants of the phases formed during the coating process. The conditions of heat tratment were 1000–1100 °C for up to 20 h.

Most of the nickel-base alloys listed in Table 2 were aluminized and chromium-aluminized by commercially available pack cementation techniques—coated by DEW (Krefeld), ONERA (Paris) and FIAT (Turin). In addition, some alloys were coated by a 'slurry process', in which the aluminium (and chromium) and the activator were fixed on the substrate surface with the aid of a liquid organic binder such as nitrocellulose lacquer. Then the samples were heat treated in a vacuum at 10^{-4} torr to obtain the diffusion layer.

The coated nickel-base alloys were annealed for 50 h at 1100 °C in flowing argon to demonstrate how the diffusion of the aluminium into the substrate depended on its composition. An electron probe microanalyzer was used to aid in the identification of the phases formed. It also proved useful for determining the thicknesses of the phases formed.

DEPENDENCE OF THE LAYER STRUCTURE ON THE COMPOSITION OF THE BASE MATERIAL

Figure 2 shows three aluminized alloys of different composition. The relatively high Ni content of the austenitic steel Thermax 4876 (see Table 1) is sufficient to form an outer NiAl layer. Under this is an extended Al-containing solid solution zone which also contains embedded precipitates of nickel aluminides.

The outer layer of aluminized nickel-base alloys also consists of NiAl (see Fig. 2b and c). The intermediate layer below is characterized by its

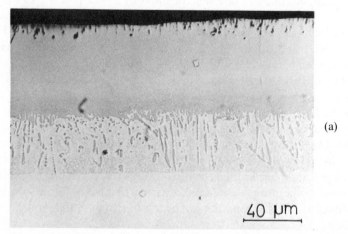

(a)

40 μm

Fig. 2.   Cross-sections of aluminized alloys with different compositions: Thermax 4876

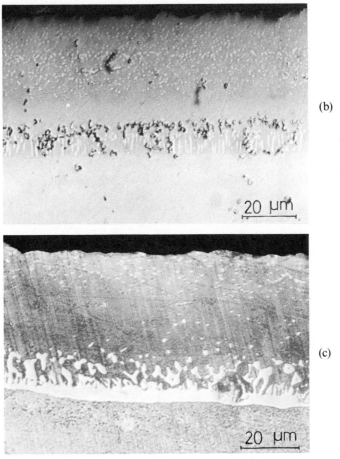

(b)

(c)

FIG. 2—*contd.*   IN-100 (top); IN-935 (bottom)

enrichment in chromium. In the case of the low chromium nickel-base alloy IN-100, this phase, with high chromium content, precipitates in the form of lamellae embedded in a Ni$_3$Al matrix. In contrast, the chromium-rich alloy IN-935, forms a compact, continuous CrNi interlayer enriched in molybdenum.

The precipitation of the chromium-rich interlayers in aluminized nickel-base alloys is due to the low solubility of chromium (less than 5w/o at 1000°C) in the NiAl.

The aluminized samples shown in Fig. 2 have been coated in a pack

TABLE 2
Composition of nickel-based alloys

Alloy	Composition (w/o)																
	Ni	Co	Fe	Cr	Mo	Si	Mn	C	Ti	Al	Nb	W	B	Zr	Ta	V	Hf
INCONEL-601	60.5	—	bal	23	—	—	0.5	0.05	—	1.3	—	—	—	—	—	—	—
HASTELLOY X	bal	1.5	18.5	22	9.0	0.5	0.5	0.1	—	—	—	—	—	—	—	—	—
IN-935	bal	19.7	24.8	24.8	1.6	—	—	0.1	3.0	1.4	1.0	—	—	—	—	—	—
ATS 351	bal	11	2.5	19	10	—	—	0.09	3.2	1.5	—	—	0.01	—	—	—	—
ATS 340	bal	16.8	0.6	19.3	—	0.3	—	0.07	2.4	1.4	—	—	—	—	—	—	—
IN-738	bal	8.5	—	16	1.8	—	—	0.17	3.4	3.4	0.9	2.6	0.001	—	1.8	—	—
Mar M 246	bal	10	—	9.0	2.5	—	—	0.15	1.5	5.5	—	10	0.015	0.05	1.5	—	—
Mar M 002	bal	10	—	9.0	—	—	—	0.1	1.5	5.5	1.2	10	—	—	1.2	—	1.3
IN-713 C	bal	—	—	12.5	4.1	—	—	0.12	0.8	6.1	2.0	—	0.012	0.1	—	—	—
ATS 290-G	bal	1.0	1.5	13	4.5	—	—	0.12	0.8	6.0	2.0	—	0.01	0.1	—	—	—
IN-100	bal	14.9	0.4	9.9	3.0	0.03	0.03	0.16	4.8	5.6	—	—	0.018	0.06	—	1.0	—
Ni/20Cr	bal	—	0.01	19.9	—	0.04	0.02	0.055	—	—	—	—	—	—	—	—	—

mixture with only small amounts of activator; in this case the coating layer grows inwards into the base material (4–6). The residual chromium, not soluble in the NiAl, will be concentrated in the regions between the larger NiAl zones as the inward-growing NiAl fronts sweep up the chromium before themselves. In this way the Cr-rich interlayer will be formed. Hard particles such as carbides and nitrides, which are present in the base material, will also be incorporated in this interlayer.

As may be seen from these examples, the structure of the aluminized samples depends not only on the Ni content but also on the Cr content of the base material. These dependences will now be considered.

INFLUENCE OF THE NICKEL CONTENT OF THE SUBSTRATE ON THE LAYER STRUCTURE

Figure 3 shows schematically the layer structure of aluminized samples with different Ni contents, the Cr content being constant at 20–25w/o except for the samples of pure iron and the alloys with Ni contents greater than 80w/o (the composition of the alloys is given in Table 2).

Such a diagram does not have any quantitative significance, because the layer structure depends considerably on coating conditions such as the composition of the pack mixture and the heat treatment, and also on the type of coating process and the flushing gas employed. Similar remarks apply to the growth rate constants of the various phases, as will be seen

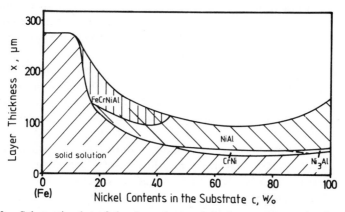

FIG. 3. Schematic plot of the dependence of the layer structure in aluminized alloys on the Ni content of the base material (Cr contents const. 20–25 w/o)

later. Therefore, it is valid only to compare results for samples which have been coated under the same conditions as those shown here.

When pure iron or Remanit 4301 containing 10w/o Ni are coated by the pack cementation used in our experiments (pack mixture: 20w/o Al, 3w/o NH$_4$ Cl, bal. Al$_2$O$_3$) only very extended solid solution layers, of nearly the same thickness, are observed. Intermetallic phases richer in aluminium can only be obtained by increasing the proportion of aluminium and activator in the pack mixture. But when base materials with Ni contents greater than 15–20w/o are used, a two-phase aluminide layer is formed consisting of an outer (Fe, Cr, Ni) Al phase and an intermediate NiAl zone. The Al-containing solid solution zone below is considerably smaller. A further increase of the Ni content in the substrate leads to a broadening of the intermediate NiAl zone, until the outer coating consists only of NiAl (Ni > 40–50w/o). An associated spontaneous formation of the CrNi interlayer takes place. This CrNi intermediate layer is replaced by a Ni$_3$Al phase, when base materials with very high Ni contents and therefore lower Cr contents are used.

Figure 4 exhibits the temperature dependence of the growth of the NiAl and solid solution phases in the alloys listed in Tables 1 and 2. Their behaviour is also described in Fig. 3.

The growth rate constants shown in Fig. 4 characterize the growth of the different phases during the coating process. A parabolic growth rate was found, which can be described by the formula $x^2 = kt$, where x is the layer thickness (cm), k the growth rate constant (cm^2/s), and t the duration of the coating process (s).

The growth rate of the protecting NiAl phase rises with increasing Ni content in the base material. Therefore, this phase has the lowest growth rate in the case of austenitic steel with 20w/o nickel, but when pure nickel is used the growth rate of the NiAl phase is the fastest.

The growth rate constants of the solid solution shown on the right-hand side of Fig. 4 give a measure of the aluminium loss from the protective NiAl layer by diffusion, and therefore also give a measure of the embrittlement of the base material. The higher the growth rate constant, the higher is the loss of aluminium.

The growth of the Al-containing solid solution is fastest with alloys which form only solid solution layers (iron and Remanit 4301). The growth rate decreases with increasing Ni content of the substrate and has the lowest value with the nickel-base alloy with CrNi interlayer; in pure nickel the solid solution grows faster again. There are several reasons responsible for these different growth rates.

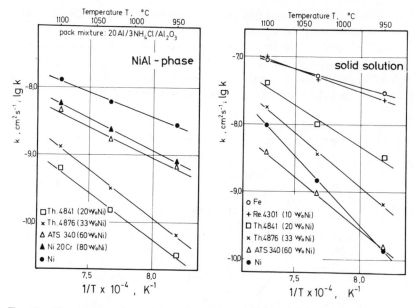

FIG. 4. Temperature dependence of the growth of the NiAl phase (left) and the solid solution zone (right) in aluminized alloys with various Ni contents and constant Cr contents between 20 and 25 w/o

The growth of the solid solution may be reduced by removal of the inward-diffusing aluminium, due to the formation of an outer aluminide-phase rich in aluminium. The diffusion in the solid solution is also influenced by its composition and by the associated metal activities. In the case of aluminized nickel-base alloys the interdiffusion of the elements between the NiAl layer and the base materials is additionally hindered by the precipitation of the CrNi phase at the β-NiAl/γ-solid solution interface; that means the CrNi interlayers have a kind of barrier effect against aluminium diffusion into the substrate.

DEPENDENCE OF THE DIFFUSION IN ALUMINIDE COATINGS ON THE CHROMIUM CONTENTS IN NICKEL-BASE ALLOYS

The barrier effect of the CrNi interlayer becomes more evident if only nickel-base alloys are coated and the Cr content in the base material is varied (Fig. 5). (The nickel-base alloys used here have been aluminized by a pack cementation process without any additions of activator (6). Of course, another mechanism of layer formation occurs, e.g. the outer NiAl layer grows outwards from the original substrate surface and the growth rates of the different phases are smaller than those coated in pack mixtures with activator, but the general layer structure is very similar.)

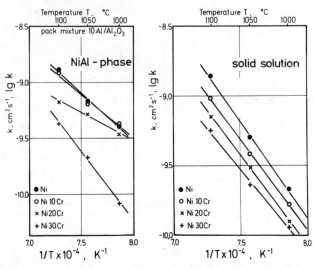

FIG. 5. Temperature dependence of the growth of the NiAl phase (left) and the solid solution zone (right) in aluminized Ni-base alloys with different Cr contents (from Meinhardt, ref. 6)

With pure nickel, which is able to form only a Ni_3Al phase as interlayer, the growth of the solid solution zone is the fastest (Fig. 5, right). With increasing Cr content in the base material the growth rate decreases continuously, providing an indication of the barrier effect of the interlayer.

The CrNi interlayers, however, also act as diffusion barriers against Ni diffusion in the nickel aluminide. This is demonstrated by the decrease of the NiAl-layer thickness with increasing Cr content in the substrate. Simultaneously with the decreasing NiAl-layer thickness, an increase of the aluminium concentration in the NiAl can be observed. This increase of the aluminium content can lead to the formation of an outer Ni_2Al_3 phase when using chromium-rich nickel base alloys (Ni/20Cr, Ni/30Cr), but in the case of Ni/10Cr and pure nickel only an outer NiAl layer is observed under the same coating conditions.

FORMATION OF INTERLAYERS IN ALUMINIDE COATINGS
ON COMMERCIAL NICKEL-BASE ALLOYS

In contrast to the Ni/Cr-model alloys, technical nickel-base alloys
contain further components besides chromium, some of which take part in
the formation of the interlayer in an analogous way to chromium. For a
simplified model the tendency of the different elements to participate in the
formation of the barrier layer can be characterized by their solubility in the
nickel-rich NiAl. That means the lower the solubility of the element
considered in the NiAl, the easier it will be pushed into the interlayer. The
normal alloying additions to the nickel-base alloys can be subdivided into
three groups:

Increasing solubility in nickel-rich NiAl

$$\longleftarrow$$

Fe, Co, Pt   Ti, Cr, Ta, W   Nb, Mo

$$\longrightarrow$$

Increasing tendency to form diffusion barriers

The left group (Fe, Co, Pt) represents those elements with the highest
solubility in NiAl. Their concentration in the NiAl layer is nearly as high as
in the substrates.

Of the elements in this list, platinum is not of course widely used as an
alloying component. Nevertheless, some years ago electrolytic deposition
was tried before the aluminization, with the intention of providing a
diffusion barrier (7). But platinum aluminides show a very high solubility in
the NiAl and for this reason a platinum solubility of up to 30w/o could be
found. So the platinum is not enriched in the interlayer. Nevertheless,
platinum can hinder the aluminium diffusion from the NiAl layer, because
it diffuses simultaneously with the aluminium and thus alters other diffusion
parameters such as the activities of the various alloy components. Still more
important seems to be its influence in enhancing the adhesion of the
alumina to the NiAl layer (8).

All the elements of the second group (Ti, Cr, Ta, W) exhibit an average
solubility of about 4w/o in the NiAl at 1000 °C. Their concentration in the
interlayer, therefore, will be approximately proportional to their
concentration in the base material.

Niobium and molybdenum exhibit the lowest solubility in NiAl, less than
1 %. So nickel-base alloys like Hastelloy X or René 41, which have in
addition to a high Cr content a relatively high Mo content, form extremely

compact CrMo(Ni)-barrier layers. In contrast to the situation with molybdenum, niobium is found in the form of carbides dispersed throughout the barrier layer.

Figure 6 shows the correlation between the effectiveness of the diffusion barrier and the concentration of the barrier-forming elements in the substrate. On the left, the thickness of the interlayers formed in aluminized

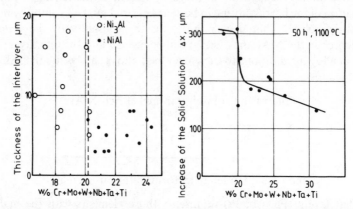

FIG. 6.   Influence of the composition of Ni-base alloys on the barrier effect of interlayers. Left: dependence of the nature and the extent of the interlayer on the total content of Cr + Mo + W + Nb + Ta + Ti in the substrate (from Fleetwood, ref. 9). Right: dependence of the increase of thickness of the solid solution zone due to the inward-diffusing aluminium (after 50 h at 1100 °C) on the total content of Cr + Mo + W + Nb + Ta + Ti in the substrate

nickel-base alloys are plotted against the total contents of chromium, molybdenum, tungsten, niobium, tantalum and titanium in these alloys. Fleetwood (9) has coated specially prepared model alloys with high γ'-Ni$_3$ (Al, Ti) contents, Cr contents between 3 and 10w/o and W contents between 7 and 20w/o. Niobium, tantalum and molybdenum were present in lower concentrations. The author gave no details of the composition of his pack mixture.

It can be seen that the matrix of the interlayer can consist of Ni$_3$Al or of NiAl. The fields where the different aluminide phases can exist are limited by critical concentrations in the base materials of the elements forming diffusion barriers. If their total concentration is less than 20w/o, the matrix of the interlayer consists of Ni$_3$Al, but at higher concentrations the matrix is based on NiAl. The average thickness of the interlayer containing Ni$_3$Al is greater than that of NiAl. It is noteworthy that the quantity of aluminium in the two types of interlayer is virtually the same, i.e. equivalent to ~1.2–1.5 μm of pure aluminium. That means that when there is more than 20w/o of diffusion barrier-forming elements in the substrate, the inward-diffusing aluminium will be retained to a greater extent in the interlayer.

A similar result is represented on the right-hand side of Fig. 6. Here the thickness of the solid solution zone formed by the aluminium diffusing inward from the outer layer (after 50 h at 1100 °C) is plotted against the total content of the elements forming diffusion barriers. The base materials were representative types of nickel-base alloys (see Table 2), which were aluminized either in our own experiments or commercially (DEW, FIAT) from aluminium donor phases activated by halides. The increase of solid solution thickness becomes smaller with increasing concentrations of barrier-forming elements in the base material, but there is a definite discontinuity in the relationship at about 20w/o. Now some remarks on the influence of the aluminium which is already present in the substrate before the aluminization. On average, the aluminium-rich nickel-base alloys exhibited a greater increase in the thickness of the solid solution interlayer than low-aluminium and aluminium-free alloys. This depends on the low capacity of the aluminium-rich solid solution to dissolve the additional inward-diffusing aluminium; in consequence a greater depth of alloy is affected. In contrast a relatively low increase in thickness of the solid solution zone in the aluminium-free alloy Ni/20Cr is observed.

CHROMIUM ALUMINIZATION

These diffusion studies indicate the possibility of enhancing the formation of diffusion barriers in alloys. This should be achieved by enrichment of the substrate surface with one of the elements forming diffusion barriers. Figure 7 shows scanning electron micrographs and characteristic X-ray (K_α) distributions of the different elements (Al, Cr, Ni, Fe) in a chromium-aluminized nickel-base alloy (IN-601), providing an example of the incorporation, in the coating, of an element which can form diffusion barriers. In this case the substrate surface is enriched with chromium before or during the aluminizing process.

The layer structure of chromium-aluminized nickel-base alloys is practically the same as that obtained by aluminization: the outer part of the layer again consists of NiAl containing CrNi precipitates as the principal phase. The additional chromium diffusion also causes the formation of a more compact and continuous CrNi-diffusion barrier.

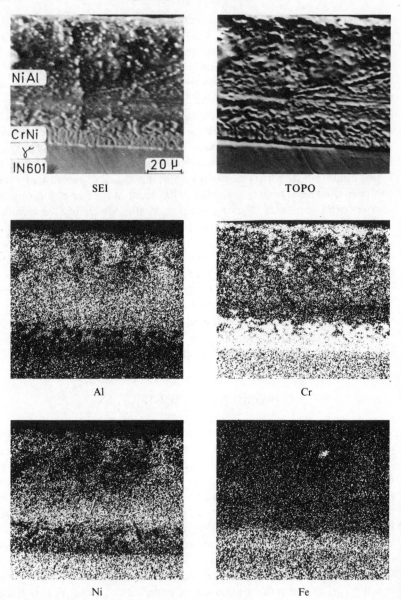

FIG. 7.   Chromium-aluminized IN-601. Scanning pictures, obtained by secondary electrons (SEI), back-scattered electrons (TOPO) and the characteristic X-rays of the elements Al, Cr, Ni, Fe

In the literature, chromium-aluminized specimens are described as being more oxidation resistant than the corresponding aluminized alloys. In particular, their increased resistance against hot corrosion caused by sulphur is cited (10). The enhanced barrier effect of the interlayer and the associated changed diffusion behaviour is referred to only in our own papers (4, 5).

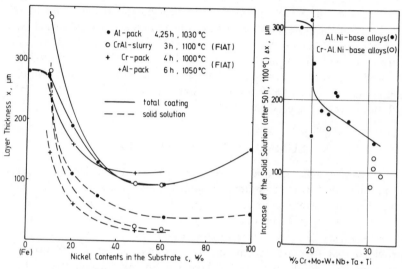

FIG. 8. Dependence of layer structure and diffusion behaviour of chromium-aluminized alloys on the composition of the base material. Left: comparison of the layer structure of our own aluminized samples with those of commercial chromium-aluminized ones (FIAT). Right: comparison of the increases of thickness of the solid solution zone of aluminized and chromium-aluminized Ni-base alloys after heat treatment of 50 h at 1100 °C

On the left-hand side of Fig. 8 the layer structure thickness of aluminized and chromium-aluminized alloys are plotted against the Ni contents of the base materials. To get a better survey only the measured thicknesses of the total layer and of the solid solution interlayer are plotted in Fig. 8. The chromium-aluminized samples were coated not only by simultaneous deposition of chromium and aluminium (CrAl slurry) but also by a chromization treatment followed by an aluminization treatment (Cr pack + Al pack). With nickel-base alloys, both methods of chromium-aluminization result in an increased formation of the CrNi(Mo) diffusion barrier in comparison with simple aluminization; this is demonstrated clearly by a decreased formation of the solid solution, although the thickness of the total layer is nearly the same. With the iron-base alloys, which are not able to form diffusion barriers, the formation of the solid solution is restricted due to the additional Cr diffusion.

However, these results alone do not provide sufficient evidence to prove the increased formation of the interlayer in the case of chromium-aluminizing, because one can obtain very similar layer structures by variations of the simple aluminization process. So the increase in the thickness of the solid solution zone of chromium-aluminized nickel-base alloys after a heat treatment of 50 h at 1100 °C was compared with that for correspondingly heat treated aluminized samples (Fig. 8, right). This showed that, due to the enrichment of the substrate surface with chromium, the formation of a barrier layer is enhanced, so the growth of the solid solution layer is reduced by 10–40%.

SUMMARY AND CONCLUSION

In contrast to the situation with iron-base alloys, nickel-base alloys spontaneously form chromium-rich intermediate layers during aluminizing. These interlayers act as barriers against the diffusion of the aluminium from the outer NiAl layer into the solid solution zone. Besides chromium, other elements can participate in the formation of the diffusion barrier, provided that they are present in the base material at a concentration exceeding their solubility in the NiAl formed. So only chromium, tungsten, titanium, tantalum, niobium and molybdenum were found to be able to form the diffusion barrier. The higher the concentration of these elements in the base material, the better is the barrier effect. This explains the clearly observed decrease in the diffusion of aluminium into the substrate when the total contents of Cr + W + Ti + Ta + Nb + Mo are more than 20w/o.

For nickel-base alloys with low contents of elements able to form diffusion barriers, it is proposed that the surface of the substrate should be enriched with chromium before or during the aluminization (chromium-aluminization). This leads to chromium-rich and therefore relatively oxidation-resistant interlayers, which have in addition an increased diffusion barrier effect. Those layers can be improved further by modification of the chromium-aluminization technique. So the oxidation resistance will further be increased by additional incorporation of platinum, rhodium or rare-earth elements into the layer. The main effect of these elements is probably the increased adhesion of the alumina formed during oxidation.

REFERENCES

1. J. Huminik (1963). *High Temperature Inorganic Coatings*, Reinhold, New York.
2. Alonized Steels, Information from the ALON Processing Inc., Pittsburgh 38, Pa (1976).
3. W. Betz, H. Huff and W. Track (1976). *Z. Werkstoff.*, **7**, 161–6.
4. E. Fitzer and H.-J. Mäurer (1977). *Z. Werkstoff.*, **8**, 112–19.
5. E. Fitzer and H.-J. Mäurer (1977/78). *High Temperature–High Pressure*, in press.
6. H. Meinhardt (1974). Diss. Universität Karlsruhe.
7. G. Lehnert and H. Meinhardt (1970). Proc. Mtg. 'Verbundwerkstoffe' (DGM, Konstanz, Germany), 281–95.
8. E. J. Felten and F. S. Pettit (1976). *Oxid. Met.*, **10**(3), 189–223.
9. M. J. Fleetwood (1970). *J. Inst. Metals.*, **98**, 1–7.
10. P. Galmiche (1969). *Corrosion*, **17**(4), 185–91.

DISCUSSION

P. Felix (BBC, Baden, Switzerland): My question is addressed to Dr Mäurer and Dr Betz. Can the differences observed at MTU between Al and CrAl coatings be explained by the formation of diffusion barriers, as described by Mäurer, or merely by differences in the heat treatment?

Reply: I do not believe that the heat treatment during the coating process produces the different mechanical behaviour of aluminized and chromium-aluminized samples, because normally very similar coating conditions are used. But, equally, I do not think that possible improvements in the effectiveness of the diffusion barrier could have such a strong influence in the case of chromium-aluminizing, because the differences shown in Fig. 8 are too small. A more important reason may result from the different layer structures. The chromium-aluminized specimens tested at MTU were prepared in a pack mixture with 'low-Al activity'. In this situation the outer NiAl phase grows outwards from the original substrate surface and will normally form a nickel-rich NiAl. But when aluminizing by halide activation the total layer grows into the substrate, thus reducing the cross-section of the high strength material. Nevertheless, an aluminium-rich NiAl layer (or Ni_2Al_3) will usually be formed, which is more brittle than the nickel-rich NiAl of comparable thickness formed by chromium-aluminizing. But in the current state of the experimental programme it is not possible to give an absolute judgement on what is the real reason for the different behaviour of aluminized and chromium-aluminized samples. In co-operation with IMM (Lehrstuhl für Metallurgie und Metallkunde, München), we are now trying to clarify this phenomenon.

17

Influence of the Mode of Formation on the Oxidation and Corrosion Behaviour of NiAl-type Protective Coatings

R. Pichoir

ONERA, Chatillon, France

ABSTRACT

The nickel-base superalloys used in the aeronautical industry are often protected by NiAl-type coatings against oxidizing and corrosive environments. Two types of technique ('low activity' and 'high activity') are generally employed in order to form the protective coatings. The specific characteristics of each technique are discussed while referring to the case of pure nickel on the one hand and to a superalloy on the other hand.

Some examples of practical interest are given, demonstrating that the hot corrosion and oxidation behaviour of NiAl-type coatings is strongly dependent on the aluminizing method. Finally it is stressed that a predeposit of suitable composition prior to aluminizing will make it possible to better exploit the inherent potential offered by the various aluminizing techniques.

INTRODUCTION

There are various techniques for obtaining NiAl-type aluminide protective coatings on nickel-base superalloys destined for certain high temperature applications in aircraft gas turbines. Preferential diffusion of nickel or aluminium occurs in the different layers which are formed during the heat cycles associated with these techniques (1, 2).

The various modes of coating formation may be grouped into two categories:

(1)   Direct formation of a NiAl compound in a single thermal cycle during the coating operation, involving preferential diffusion of

271

nickel. The typical temperature range of the heat treatment is 1000–1100 °C.

(2) Formation of an initial layer of a Ni_2Al_3 compound with preferential diffusion of aluminium; subsequent transformation of the layer into NiAl through diffusion annealing, with preferential diffusion of nickel. These latter-type processes comprise two heat cycles: an aluminizing treatment performed generally within the temperature range 700–850 °C, and subsequent diffusion annealing in argon at temperatures between 1000 and 1100 °C.

In fact, there is another type of process which involves a single heat cycle during which, essentially, a Ni_2Al_3 layer is formed first, followed by the formation of the NiAl coating. Therefore, this process has a strong resemblance to those in the second category mentioned above, but its advantages and disadvantages are not discussed here.

The aluminizing processes are frequently of the pack cementation type. The cement is generally in the powder form and consists of: aluminium-rich metallic powders; a halide to achieve aluminium transport from the cement toward the component to be protected; and an inert diluent—powders of stable oxides.

In these processes, cements consisting of more or less aluminium-rich metallic powders are used in order to obtain layers, more or less rich in aluminium. In most cases, either Ti–Al, Ni–Al or Cr–Al, etc., alloy powders are utilized.

By aluminizing Ni-base superalloys, depending upon the aluminium content of the powders, one can obtain either directly NiAl-type layers (first category processes) or Ni_2Al_3-type layers subsequently transformed to NiAl coatings by diffusion (second category processes).

In both types of process the quantity of aluminium which is necessary to form the coating is very small compared with the amount contained in the cement. Therefore, the composition of the metallic powders is hardly modified during the aluminizing treatment.

PROCESSES OF THE FIRST CATEGORY: DIRECT FORMATION OF NiAl COATING

Aluminizing techniques which allow direct formation of the NiAl compound are the so-called 'low activity processes'. If the aluminium content of the NiAl formed is either near-stoichiometric or under-stoichiometric, it is primarily the diffusion of nickel which occurs. The

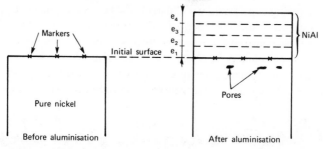

FIG. 1. Aluminization of pure nickel (low activity process; diffusion of Ni through NiAl): first e_1 is formed, then $e_2, e_3, \ldots,$ etc.

consequences of this preferential diffusion are described in the following two cases: the simple case of pure nickel and the case of nickel-base superalloys.

Pure Nickel

The major consequences of the diffusion of nickel in NiAl may be briefly described as follows (Fig. 1):†

(1) The initial surface of the nickel specimen corresponds to the NiAl–Ni interface; the stable particles such as oxides and diffusion markers which can be embedded in the surface are located at this interface.

(2) Kirkendall porosities may occur; their importance will depend upon the chosen experimental conditions, i.e. temperature and time of aluminizing treatment. In nickel, these pores appear in the vicinity of the nickel/aluminide interface (Fig. 2).

(3) Since it is a cementation-type process, the various powders constituting the cement may become embedded in the coating during the aluminizing operation. The powders of the inert diluent are observed throughout the coating, whereas the metallic powders appear almost invariably only in the superficial zone (Fig. 2) because they tend to dissolve in the NiAl compound with increasing time; the internal zones are precisely the ones formed at the beginning of the heat treatment (see Fig. 1). The gradual dissolution of powder inclusions results in a modification of the overall composition of the NiAl layer.

† The initial surface (or initial limit) is drawn at the same position in every figure showing the different steps of the layer formation.

The following experiment will illustrate the last consequence (trapping of cement powders): during the single aluminizing operation thin sheets of pure nickel were entirely transformed into NiAl in three different ways (Fig. 3a): sheet 1, completely in contact with the cement; sheet 2, one face in contact with the cement and no contact with the other; and sheet 3, no contact with the cement, sheet completely aluminized in the vapour phase.

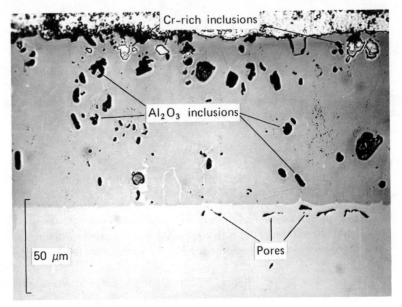

FIG. 2.   Aluminization of pure nickel by 'chromaluminizing' process (low activity)

The cement employed in this particular case has the same composition as the one used in the 'chromaluminizing' process (3); it essentially consists of metallic powders of 85Cr/15Al alloy and alumina powders (inert diluent). A small quantity of ammonium chloride is added to these powders in order to ensure the transport of aluminium from the cement to the specimens.

The metallographic observations of the NiAl sheets thus obtained clearly show the alumina particle inclusions transported to those faces which are in contact with the cement (sheets 1 and 2, Fig. 3b). The metallic particles are also embedded but are primarily observed close to the surface (Fig. 4), since they have dissolved in the central part of the NiAl sheet (see Fig. 1).

The chemical analysis of these three sheets of NiAl (Fig. 3c) confirms the presence of alumina inclusions as well as the chromium enrichment of the

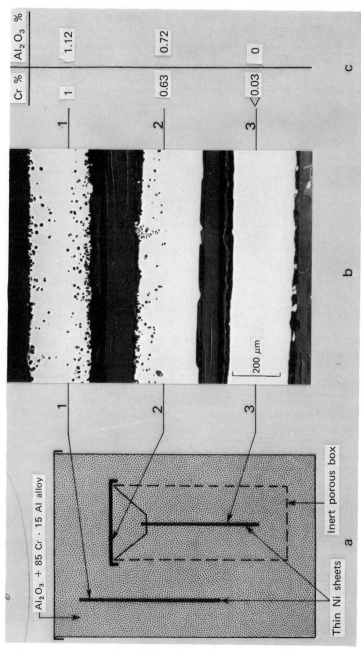

FIG. 3. Aluminization of thin pure nickel sheets by 'chromaluminizing' process (low activity): (a) schematic drawing of the device; (b) nickel sheets after complete transformation into NiAl; (c) chemical analysis of each sheet

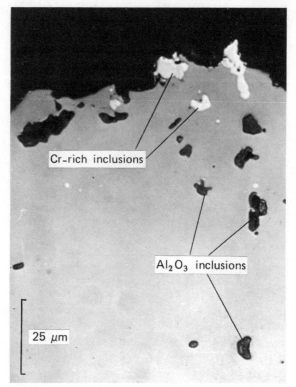

FIG. 4. Thin nickel sheet completely transformed into NiAl by 'chromaluminiz-ing' process: metallic powder inclusions are observed only in the vicinity of the surface

NiAl compound when the specimens are in contact with the cement. In addition, the following remarks should be made:

(1) A comparison of the results of chemical analysis for chromium and alumina, in the case of sheets 1 and 2, shows that the embedding of the cement particles is made easier by a good contact between the cement and the specimen. In fact, during the aluminizing operation the thin sheets are deformed and hence it is easier for the cement to maintain a better contact with the upper face of the horizontal sheet (sheet 2), as compared with the case of the vertical sheet 1. Consequently, in the case of sheet 1, the chromium and alumina contents are always significantly lower than twice the values obtained from sheet 2.

(2) Hence, the chromium content is found to be less than 0.03% in sheet 3; the transport of this element through chromium vapour halides as well as by direct thermal evaporation is considered to be negligible. For the specimens in direct contact with the cement, the observed chromium enrichment therefore essentially occurs through the metallic powder inclusions. However, it is quite possible that some chromium transfer may take place by solid-state diffusion between the non-included powders and the specimen.

Nickel-base Superalloys

The nickel-base superalloys generally contain significant amounts of aluminium (3–6%) and therefore, compared with the case of pure nickel, the consequences of preferential diffusion of Ni in NiAl could be different:

(1) The initial surface of the specimen is no longer at the coating/substrate interface but is now located inside the coating. One may therefore schematically represent the coating as two zones on either side of the initial surface (Fig. 5):

  (a) An external zone of NiAl formed in the same way as the whole layer in the case of pure nickel, i.e. by diffusion of nickel in NiAl.

  (b) An internal zone formed from the underlying zone of the alloy; owing to the diffusion of nickel toward the coating, the underlying zone will become denuded in nickel and rich in the various other elements constituting the alloy, especially aluminium. This particular zone transforms into NiAl-containing precipitates formed from those elements of the alloy that cannot be completely dissolved in NiAl (Fig. 6).

  Any foreign particles attached to the specimen surface prior to aluminizing remain at their original spatial position and thus eventually appear in the interior of the coating, marking the position of the initial surface. These particles are dangerous because they may favour spalling of the protective coating during the thermal cycling of the components. Consequently an appropriate cleaning of the surface, prior to aluminizing, is highly recommended.

(2) The Kirkendall porosities are generally absent and this may be explained, at least partially, by the fact that for the same coating thickness there is less nickel transport in the case of a superalloy than in the case of pure nickel. Indeed, for superalloys, the nickel transport corresponds to the quantity necessary to form the

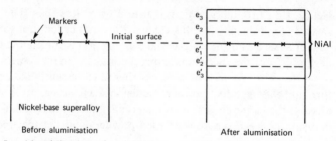

FIG. 5. Aluminization of a nickel-base superalloy (low activity process; diffusion of nickel through NiAl): first e_1 and e_1' are formed at the same time, then e_2 and $e_2', \ldots$, etc.

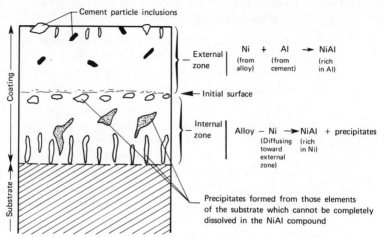

FIG. 6. Schematic illustration of a NiAl-type coating structure on nickel-base superalloys (low activity process)

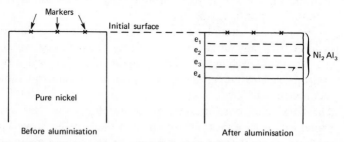

FIG. 7. High activity process—formation of a Ni_2Al_3 layer in the case of pure nickel (diffusion of aluminium through Ni_2Al_3): first e_1 is formed, then $e_2, e_3, \ldots$, etc.

external zone alone, whereas for pure nickel it corresponds to the formation of the whole coating.

(3) The powders constituting the cement can only be embedded in the external zone of the coating (Fig. 6). For the particular case of metallic powder inclusions, it will be shown later that their presence can considerably modify the corrosion behaviour of the component.

PROCESSES OF THE SECOND CATEGORY: FORMATION OF A Ni_2Al_3 LAYER FOLLOWED BY HIGH TEMPERATURE DIFFUSION ANNEALING TO OBTAIN A NiAl-TYPE COATING

Formation of a Layer of Ni_2Al_3 Compound

The aluminizing techniques associated with the formation of Ni_2Al_3 are known as 'high activity processes'. In this compound, it is principally aluminium which diffuses and, in the previously mentioned two cases, the consequences are as follows.

Pure nickel (*Fig. 7*)

(1) The initial surface of the specimen corresponds to the surface of the Ni_2Al_3; the stable particles adhering to the specimen surface will therefore be located at the surface of this layer after the aluminizing treatment.

(2) There is no Kirkendall effect in the specimen.

(3) The powders of the cement cannot be embedded in the Ni_2Al_3 layer; the composition of this layer may however be superficially modified either by solid-state diffusion between the metallic powders and the layer or by metallic halide vapour phase diffusion.

Superalloys (*Fig. 8*)

All the previous remarks apply to the superalloys. It should also be noted that those elements of the substrate that are not completely soluble in Ni_2Al_3, precipitate in this compound. Moreover, the stable particles such as MC-type carbides, already existing in the alloy region transformed by aluminizing, remain embedded in the Ni_2Al_3 layer.

Diffusion Annealing

The annealing treatment, preferably carried out in an argon atmosphere, transforms the Ni_2Al_3 layer into a NiAl coating.

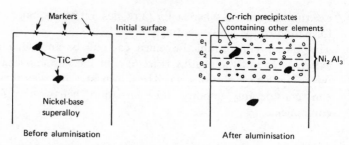

FIG. 8.   High activity process—formation of a Ni_2Al_3 layer in the case of a nickel-base superalloy (diffusion of aluminium through Ni_2Al_3): first e_1 is formed, then $e_2, e_3, \ldots$, etc.

Pure nickel

In the case of pure nickel, diffusion annealing leads to the formation of NiAl, while the Ni_2Al_3 compound gradually disappears (Fig. 9). During this annealing process there is preferential diffusion of nickel across a portion of the NiAl compound which is being formed and therefore, depending upon the annealing conditions, Kirkendall porosities may appear. Nevertheless, the extent of pores is limited by the fact that a significant portion of the NiAl compound is obtained through the loss of aluminium by the initial Ni_2Al_3 layer.

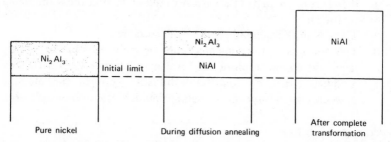

FIG. 9.   High activity process: formation of a NiAl coating from a Ni_2Al_3 layer, in the case of pure nickel, by diffusion annealing

Superalloys

The NiAl coating obtained by interdiffusion between a layer of Ni_2Al_3 compound and the superalloy substrate consists of three regions (Fig. 10): (1) an external zone almost as thick as the initial Ni_2Al_3 layer—this zone contains various precipitates already existing in the initial layer; (2) a central region, devoid of precipitates; and (3) an internal zone containing

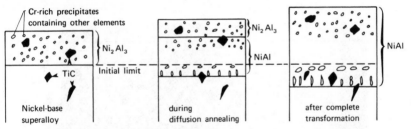

FIG. 10.   High activity process: formation of a NiAl coating from a Ni_2Al_3 layer, in the case of a nickel-base superalloy, by diffusion annealing

precipitates similar to those observed in the internal zone of coatings obtained after a 'low activity' aluminizing treatment (see Fig. 6).

The last two zones may in fact be considered to be equivalent to a 'low activity' coating, the initial Ni_2Al_3 layer here playing the role of cement. It is therefore the external zone, containing precipitates formed in the initial Ni_2Al_3 layer, which primarily distinguishes the high activity NiAl coatings from the low activity ones. The nature of the alloying elements of the substrate, constituting these precipitates, may favourably or adversely affect the oxidation and corrosion behaviour of the coating.

It should also be noted that the cement particle inclusions are not observed in the coatings: the effect of these particles is dependent on their nature. Moreover, since the initial surface of the alloys corresponds to the surface of the coating (Fig. 10), there is no danger of including the superficial oxide particles (present on the substrate surface) into the coating.

EXAMPLES

Three examples are given as follows:

(1)   This example concerns the process of inclusion of powders, constituting the cement, in low activity aluminizing techniques; it illustrates the importance of chromium-rich inclusions on the corrosion behaviour of coatings.

(2)   This example refers to high activity processes; it illustrates the role of various precipitates existing in the external zone (corresponding to the initial layer of Ni_2Al_3) of the coating.

(3)   This example shows the usefulness of achieving the correct layer of metal or alloy on the components prior to the application of either low or high activity coating techniques.

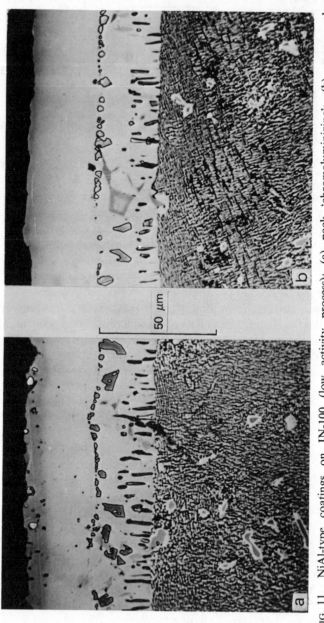

50 μm

Fig. 11. NiAl-type coatings on IN-100 (low activity process): (a) pack 'chromaluminizing'; (b) vapour phase 'chromaluminizing'

'Chromaluminizing' of IN-100 Superalloy

Samples of IN-100 alloy were protected by using the cement appropriate for the chromaluminizing process. Some samples were put directly in contact with the cement, while others were isolated from it by a porous wall (see Fig. 3a) and hence aluminized in the vapour phase.

Figures 11(a) and 11(b) show metallographic cross-sections of each type of sample. In Fig. 11(a), some particle inclusions of the cement can be observed in the vicinity of the surface, whereas the coating shown in Fig. 11(b) does not contain any of those inclusions. Moreover, the volume fraction of precipitates in the internal zone is slightly lower in the case of samples that have been aluminized in the vapour phase; most of these precipitates (either of CoCr-type or carbides) are rich in chromium—they can dissolve partially in the NiAl formed from the vapour phase since in this case the latter will not be enriched in chromium through inclusion of 85Cr/15Al cement particles. Finally, another difference should be pointed out: the grain size of NiAl in the external zone of the coatings is larger in the case of the vapour phase aluminized samples than in those obtained by pack cementation techniques.

Although both types of sample show similar oxidation behaviour, their responses to corrosion in the presence of Na_2SO_4 are very different (Figs. 12a and 12b). Samples containing chromium-rich cement particle inclusions are generally not degraded after 400 h corrosion at 900 °C under the conditions mentioned in Fig. 12. On the contrary, the samples aluminized in vapour phase are very often completely destroyed after 50 h corrosion. In fact, these samples are more or less degraded even after 25 h exposure (Fig. 12b).

High Activity Type Coatings on IN-100 Superalloy

In order to separate the two stages in the high activity processes the samples were coated as follows: high activity aluminizing at 700 °C for about 7 h in a cement containing 35 % Cr and 65 % Al; and diffusion annealing for 5 h at 1085 °C in an argon atmosphere to transform the Ni_2Al_3 layer into a coating of NiAl.

Figure 13(a) represents the metallographic structure of the Ni_2Al_3 layer obtained by the high activity aluminizing process. The following points should be noted:

(1) The absence of a NiAl diffusion zone,

(2) The presence of small precipitates rich in chromium containing some molybdenum, these two elements of the alloy not being completely soluble in the Ni_2Al_3 compound,

R. Pichoir

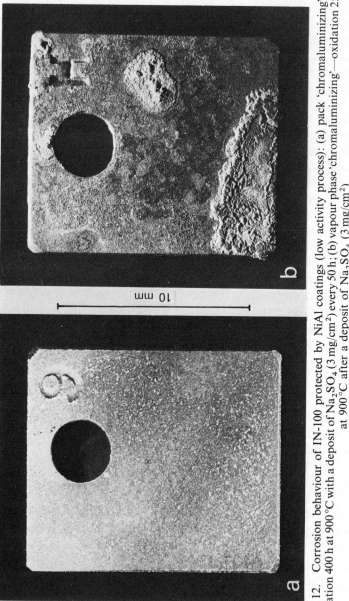

Fig. 12. Corrosion behaviour of IN-100 protected by NiAl coatings (low activity process): (a) pack 'chromaluminizing'—oxidation 400 h at 900°C with a deposit of Na$_2$SO$_4$ (3 mg/cm^2) every 50 h; (b) vapour phase 'chromaluminizing'—oxidation 25 h at 900°C after a deposit of Na$_2$SO$_4$ (3 mg/cm^2)

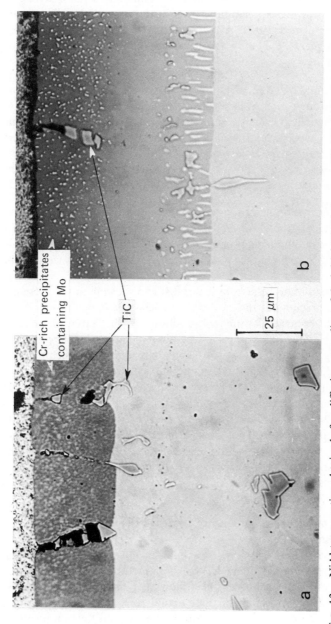

FIG. 13. NiAl-type coating obtained after diffusion annealing of the initial Ni_2Al_3 layer on IN-100 (high activity process): (a) Ni_2Al_3 layer; (b) NiAl-type coating

(3) The presence of TiC precipitates containing some molybdenum. Most of these precipitates exist in the zone of the alloy which is transformed into Ni_2Al_3—they simply grow during the aluminizing treatment; one may frequently observe some defects in their vicinity.

Figure 13b shows the metallographic structure of the NiAl coating obtained through interdiffusion of the Ni_2Al_3 layer and the underlying alloy, this metallographic structure corresponds to the description under 'Superalloys', p. 280, and is in accordance with the schematic diagram shown in Fig. 10. In the case of IN-100 alloy, and in contrast with the previously described low activity type of coatings, the behaviour of this high activity type coating can be summarized as follows:

(1) Its corrosion behaviour is satisfactory, but is generally inferior to that of the low activity type coatings obtained by the chromaluminizing process. On the other hand, this type of coating is far superior to the low activity coatings obtained by vapour phase diffusion.

(2) Its oxidation behaviour is unsatisfactory owing to the presence of TiC precipitates and the defects associated with them, and probably also because of the composition of the external zone. An accelerated oxidation is indeed observed around these precipitates (Fig. 14). Long oxide protrusions, extending along the precipitates and consisting mainly of titanium oxide, are formed (Figs. 15a and 15b). One may also observe on these illustrations that the oxide layer which is formed on the entire coating contains plenty of titanium oxide; such a layer is hardly protective. Its formation probably results from the fact that the external zone of these types of coating is rich in titanium, since it contains practically all the titanium of that portion of the alloy which had been previously transformed into Ni_2Al_3.

Usefulness of an Appropriate Alloy Deposit Prior to Low or High Activity Aluminizing

The description of the principal mechanisms governing the formation of different types of coating and the consideration above of two examples of coating, allows one to stress the following points. For both types of technique, the composition of the external zone of the coating, directly involved during corrosion or oxidation, depends on the composition of

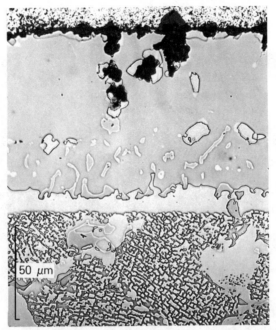

FIG. 14. NiAl-type coating on IN-100 (high activity process): oxidation 100 h at 1100 °C

the nickel-base superalloy and on the composition of the aluminium-base cement.

In low activity techniques, the elements of the substrate can diffuse across the coating being formed; consequently their concentration in the external zone of the coating is limited by their solubility in the NiAl compound.

These alloying elements may however seriously affect the corrosion or oxidation behaviour of the coating. For example, the harmful role of titanium has already been mentioned in the case of oxidation (4). The chemical composition of the cement can play a role through vapour phase transport and/or by solid-state diffusion of these various elements. The cement composition can also play an important role due to the fact that the cement particles may be entrapped (pack cementation).

In the case of high activity methods, the composition of the external zone of the coating is even more dependent upon the substrate composition: all the elements of the alloy remain confined in the initial layer of Ni_2Al_3; consequently they are present (in quantities directly corresponding to their

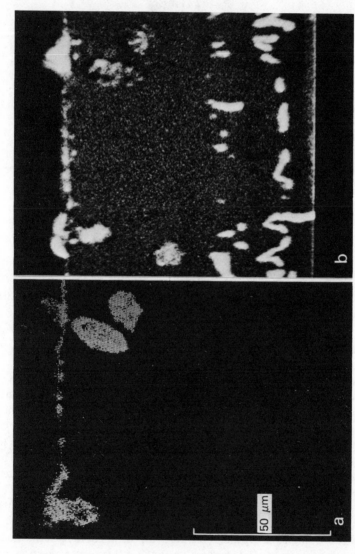

50 μm

FIG. 15.   NiAl-type coating on IN-100 (high activity process)—oxidation 100 h at 1100°C; (a) X-ray oxygen image; (b) X-ray titanium image

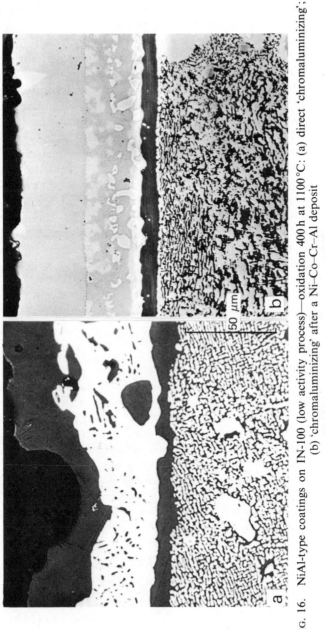

Fig. 16. NiAl-type coatings on IN-100 (low activity process)—oxidation 400 h at 1100°C: (a) direct 'chromaluminizing'; (b) 'chromaluminizing' after a Ni–Co–Cr–Al deposit

concentrations in the alloy) in the external zone of the NiAl coating which is obtained after a diffusion anneal. These elements are either in solid solution or in the form of precipitates. As far as the composition of the cement is concerned, it can play a role only through vapour phase transport and/or by solid state diffusion of the constituent elements.

Most of the troubles associated with various protective coating techniques can be avoided by first laying down a 15–30 μm thick deposit of appropriate composition prior to the aluminizing treatment. Thus, for example, a simple deposit of Ni–Cr alloy, followed by aluminizing by a high activity process, endows the coating with a superficial region rich in chromium and useful for good corrosion resistance. Similarly, a predeposit of a titanium-free 66Ni/20Co/10Cr/4 Al alloy aluminized by a low activity process allows one to obtain highly oxidation-resistant coatings (5). The behaviour of such a coating is compared to that of one obtained by direct chromaluminizing of IN-100 alloy in Figs. 16(a) and 16(b). It should be noted that the application of a prior deposit, of composition specified above, leads to good oxidation behaviour, associated in the case of IN-100 with little interaction between the coating and the alloy.

CONCLUSION

The corrosion and oxidation behaviour of coated components is strongly dependent on the aluminizing technique. The knowledge of the different mechanisms involved in each case is necessary in order to choose the aluminizing technique which should be used, taking into account the alloy composition and the conditions of utilization of the coated parts. A predeposit of an appropriate alloy allows a better use of the advantages offered by these different types of technique and thereby leads to improved coatings. Moreover, such an approach will yield NiAl-type coatings on cobalt-base superalloys and also improve considerably the protective coatings for nickel- or cobalt-base carbide-fibre-reinforced directionally solidified composites.

ACKNOWLEDGEMENTS

The author is grateful to Dr T. Khan, research scientist at ONERA, for discussions and translation of this chapter.

REFERENCES

1. G. W. Goward, D. H. Boone and C. S. Giggins (1967). *Trans. Am. Soc. Met.*, **60**, 228.
2. G. W. Goward (1970). *J. Metals* (October), 31.
3. P. Galmiche (1968). *Metals Materials*. (August), 241.
4. B. Dupré, P. Steinmetz, B. Roques and R. Pichoir (1977). *Rech. Aérosp.*, no. 4, p. 213.
5. R. Pichoir (1975). *Proc. 5th Int. Conf. on Chemical Vapor Deposition* (Eds. J. M. Blocker *et al.*, Electrochemical Society, Princeton, N.J.), p. 298.

DISCUSSION

P. Felix (BBC, Baden, Switzerland): How can you explain that particles of the pack powder are found throughout the whole cross-section of a thin specimen after coating a pure metal with Al, although only base metal diffuses outward but no Al inwards?

Reply: At a given time, t, cement particles in contact with the surface may be considered as diffusion markers. Moreover, only Ni can diffuse through NiAl already formed. Due to this fact, some of the cement particles (in contact with the surface) may become entrapped in the NiAl being formed at the above-mentioned time t (see Fig. 1). This mechanism is operative right from the beginning of the aluminizing treatment of the thin pure nickel specimen till its complete transformation into NiAl. Consequently, the alumina particles are found in the entire cross-section of the NiAl thus obtained.

In contrast, during the formation of a compound through which only aluminium diffuses, there is no possibility of entrapping cement particles. At all times, the formation of such a compound takes place at the interface between nickel and the already formed compound. This is exactly what happens during the formation of Ni_2Al_3 (see Fig. 7).

18

High Temperature Behaviour of Protective Coatings on Ni-base Superalloys

P. C. Martinengo and C. Carughi

FIAT (CRF), Orbassano, Italy

U. Ducati and G. L. Coccia

Polytechnic School of Milan, Italy

ABSTRACT

A slurry method for forming impregnation layers on Ni superalloys has been studied and optimized. CrAl, CrAlY and CrAlCe coatings have been obtained and their oxidation and hot corrosion behaviour has been observed by weight change and electrochemical tests. Y additions proved to be the most effective in enhancing the scaling properties of the CrAl coating.

INTRODUCTION

High temperature protective coatings on Ni-base superalloys are currently made by CrAl impregnation. Diffusion of the coating elements into the base alloy and spalling of the oxide layers are the usual life-limiting factors of such coatings.

An obvious way of preventing these drawbacks is to modify the coating layer composition by addition of elements which can slow down the diffusion rate and enhance the adhesion of the scale, particularly when it is subjected to thermal shocks.

In the case of superalloys, yttrium or cerium are generally known to be effective in improving oxidation and corrosion resistance.

Our research work was thus intended to check the possibility of further improving the protective properties of coatings by adding Y or Ce to CrAl-impregnated layers.

293

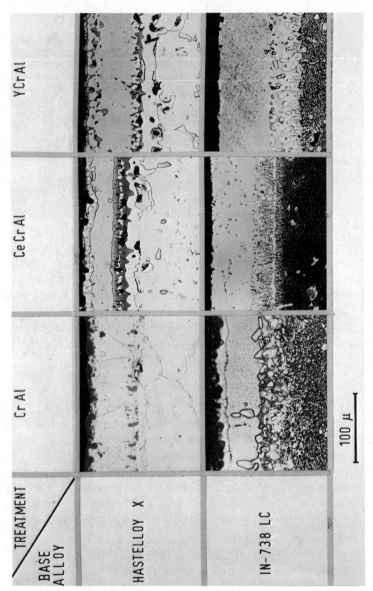

FIG. 1. Micrographs of cross-sections of the coatings

A slurry process has been developed which allows homogeneous and uniform coatings to be obtained even on surfaces of intricate shapes.

The addition of yttrium has proved to be particularly beneficial for the behaviour of the coating; the formation of yttrium needles protruding from the scale into the metal is probably the reason for this success.

The effect has been confirmed by crucible tests in fused chloride/sulphate salt mixes.

TESTS AND RESULTS

The research work was carried out on two nickel-base superalloys, Hastelloy X and IN-738, whose compositions are given in Table 1. Three types of coating, CrAl; YCrAl and CeCrAl, have been studied. Samples have been impregnated by the slurry process. Chromium, aluminium, yttrium or cerium metal powders and ammonium chloride, having a mean grain size of $\sim 5\,\mu$m have been dispersed, in an organic binder, and painted on the samples.

Heat treatment in inert gas subsequently allowed the metal to diffuse into the base alloy. By preliminary tests we optimized the conditions to get homogeneous layers. The optimum conditions are given in Table 2.

The micrographic appearance and the elemental distribution patterns of the coating layer are shown in Figs. 1 and 2, respectively.

The effect of the additions of yttrium and cerium on the diffusion rate of the elements into the bulk has been tested by heating the samples at 1000 °C for 1000 h. The results of this test are shown in Fig. 3; Ce proved to be really effective in slowing down the diffusion rate.

The evaluation of the oxidation resistance of the coatings in still air has been carried out by heating the samples in a furnace at 1100 °C and weighing them every 50 h. The specimens were extracted from the furnace, allowed to cool, ultrasonically cleaned to detach the spalled oxide and the weight change per unit surface area was determined.

Experimental results on the behaviour of Cr-aluminized, YCr-aluminized and CeCr-aluminized specimens are summarized in Fig. 4. This type of test proved to be unsuitable for discriminating clearly between the various coating process. The micrographs of the specimens after the test are presented in Fig. 5; it is possible to see that the main difference between uncoated and coated samples is that the internal oxidation is lower in the latter case; Y and Ce additions appear to further decrease the oxidation

TABLE 1
Base material compositions

Materials	Composition (w/o)										
	Cr	Ni	Co	Fe	Ti	C	Mn	Si	Al	Mo	W	others
Hastelloy X	22	47.3	1.5	18.5	—	0.1	0.5	0.5	—	9	0.6	—
IN-738	16	60	8.5	0.5	3.4	0.05	0.2	0.5	3.4	1.75	2.60	Ta: 1.75 Nb: 0.9 Zr: 0.1

TABLE 2
Coating process parameters

	1st slurry applied			2nd slurry applied			Slurry amount		Diffusion treatment	Diffusion layer thickness (μm)	
							1st slurry	2nd slurry		Hast X	IN-738
	Cr	Al	NH$_4$Cl	Binder	Al	NH$_4$Cl	Binder					
CrAl	61	18	8	13	83	4	13	75 mg/cm^2	25 mg/cm^2	1000°C 5 h	70	80
CrAlY	Cr	Al	Y	Binder	Al	NH$_4$Cl	Binder					
	64	18	9	9	83	4	13	75 mg/cm^2	25 mg/cm^2	1000°C 3 h	90	100
CrAlCe	Cr	Al	Ce	Binder	Al	NH$_4$Cl	Binder					
	64	18	9	9	83	4	13	75 mg/cm^2	25 mg/cm^2	1100°C 3 h	65	70

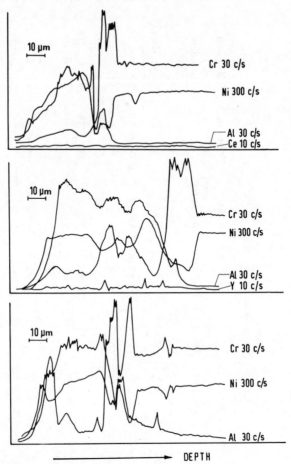

Fig. 2.   Microprobe analysis of coating layer on IN-738 LC. Vertical axes: count rate

rate, but the effect of these elements will be considered in more detail subsequently.

Thermal shock tests have been carried out by cycling the specimens between 1100 and 200 °C; samples were held for 4 min at each temperature in the furnace. Every 250 cycles the samples were examined; Fig. 6 shows plots of weight changes per unit area versus number of cycles.

In this case the difference between the Cr-aluminized and Y or CeCr-aluminized specimens is higher but the test results are independent of the rare-earth metal additions. To get more detailed information on the

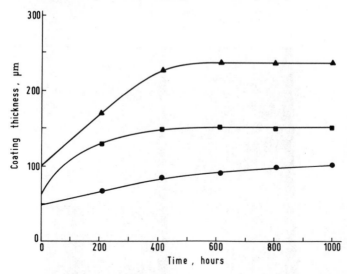

FIG. 3. Time dependence of thickness of the coating layer on IN-738 LC ($T = 1000\,°C$). ▲ YCrAl coating; ■ CrAl coating; ● CeCrAl coating

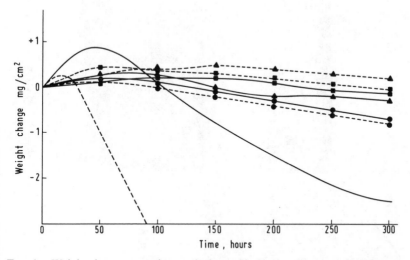

FIG. 4. Weight change per unit area during oxidation in still air at $1100\,°C$. Base material: – – – IN-738, —— Hastelloy X. ▲ YCrAl coating; ■ CeCrAl coating; ● CrAl coating

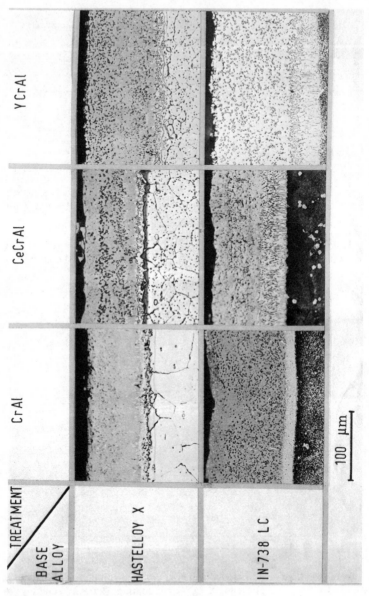

FIG. 5.   Micrographs of specimens after oxidation in air at 1100°C

oxidation products, analyses of the scale were performed. Coated specimens of IN-738 LC were oxidized at 1100 °C for 150 h and then the metal was dissolved away in a 20 % bromine–methanol solution. The oxide flakes so obtained were rinsed in pure methanol and then observed by scanning electron microscopy, transmission electron microscopy and small angle electron diffraction.

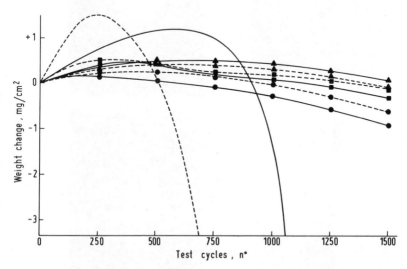

FIG. 6.   Weight change per unit area during thermal shock tests in air (1100–200 °C). Base material: - - - IN-738, —— Hastelloy X. ▲ YCrAl coating; ■ CeCrAl coating; ● CrAl coating

In the flakes from the YCr-aluminized specimens, there are many pegs protruding from the oxide layer into the metallic substrate, acting as bonds which mechanically 'key' the scale to the alloy.

Selected area diffraction (SAD) examination of the scale from the YCr-aluminized specimens shows a columnar growth of Al_2O_3 scale with their basal planes corresponding to the (00.1) crystal lattice plane and having lattice parameters $a = 4.76$ Å; $c = 13.01$ Å. They are in a matrix of spinel, probably containing both Cr and Ni; the metal to oxygen ratio has not been determined. The pegs are undoubtedly cubic yttrium crystals.

Figure 7 shows the scale analysis; the Y_2O_3 needle axes are (001) crystal axes with a measured BCC lattice parameter $a_0 = 10.77$ Å (10.58 Å is the literature value).

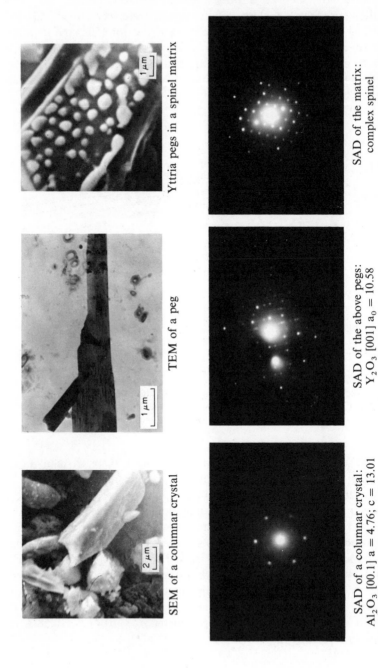

Yttria pegs in a spinel matrix

TEM of a peg

SEM of a columnar crystal

SAD of the matrix:
complex spinel

SAD of the above pegs:
Y_2O_3 [001] $a_0 = 10.58$

SAD of a columnar crystal:
Al_2O_3 [00.1] $a = 4.76$; $c = 13.01$

Fig. 7. Micrographs and SAD patterns of scale constituents after oxidation in still air at 1100 °C of an IN-738 LC sample coated with CrAlY

RESISTANCE TO HOT CORROSION

A number of conventional tests have been performed to characterize the hot corrosion behaviour of the coated alloys: rig tests, salt corrosion and thermobalance measurements.

In Fig. 8 a thermobalance measurement of the weight gain of a YCr-aluminized IN-738 LC sample is shown, corrected for weight losses due to

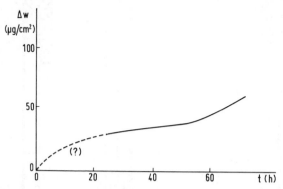

FIG. 8. Thermobalance result for weight gain of an YCr-aluminized sample of IN-738 LC (5 mg/cm^2 Na$_2$SO$_4$ + 10 % NaCl salt contamination; oxidation at $T = 920\,°C$ in still air)

evaporation. Oxidation conditions are given. As is clearly shown, the information content of the measurement is very poor: in fact we had to mark the y-axis in $\mu g/cm^2$; the total weight gain of the sample was, moreover, one order of magnitude lower than the correction for salt evaporation, as measured on an alumina rod of corresponding size. At the end of the run the sample had the appearance of a bright metal rod, locally tarnished by oxidation products.

The excellent degree of protection afforded by the coatings was proving an obstacle to characterizing their behaviour and to discriminating between the effects of the various coating procedures. Electrochemical techniques, on the other hand, appeared to offer promise as a suitable test method. By controlling the appropriate electrical variables, many corrosion phenomena may be studied at high sensitivity or even accelerated.

The cell used for the experimental work is shown in Fig. 9; the following criteria were used whenever possible: (a) careful definition of the electrode working area, (b) maintenance of a uniform current density (c.d.) distribution on the electrode and (c) imposition of working conditions as

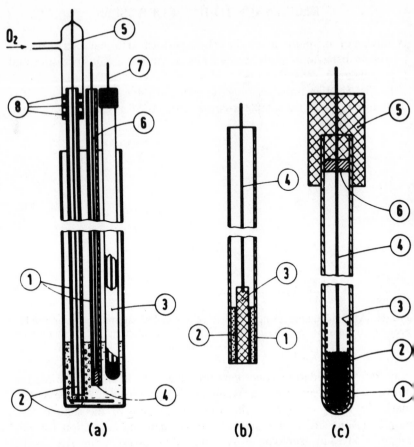

FIG. 9. Cell and electrodes used in electrochemical tests. (a) Cell: 1, sintered alumina tubes; 2, platinum; 3, stabilized zirconia reference electrode; 4, working electrode; 5, 6, 7, electrode leads; 8, gas seals. (b) Working electrode: 1, alumina tube; 2, hot-rolled BN gasket; 3, sample; 4, electrical lead wire. (c) Reference electrode: 1, stabilized zirconia tube; 2, Ni + NiO mixture; 3, Ni flash; 4, Ni wire; 5, epoxy sealing; 6, PTFE

close as possible to the temperatures and chemical environments experienced by the turbine components in service.

Measured and controlled electrical parameters were as follows: (i) current recording at steadily increasing voltage, (ii) current recording at constant voltage and (iii) voltage recording at steady current and at current transients (both single and double step).

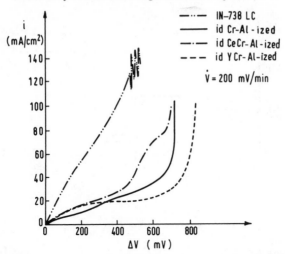

FIG. 10. Current vs. voltage plot for variously treated IN-738 LC samples in potentiodynamic sweeps. $\dot{V} = 200\,\text{mV/min}$; $T = 920\,°C$; $Na_2SO_4 + 10\%$ NaCl molten salt mixture

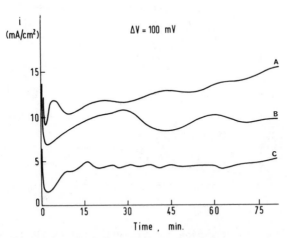

FIG. 11. Current vs. time plot for variously treated IN-738 LC samples in potentiostatic conditions. $\Delta V = 100\,\text{mV}$; $T = 920\,°C$; $Na_2SO_4 + 10\%$ NaCl molten salt mixture. A = CrAl coating; B = CeCrAl coating; C = YCrAl coating

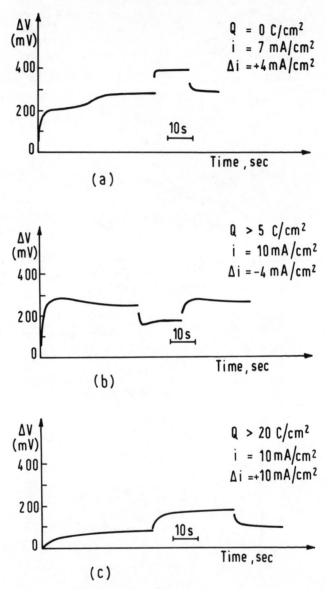

FIG. 12. Voltage vs. time plots for YCr-aluminized IN-738 LC samples in galvanostatic conditions: (a) at the beginning of the run; (b) after $5\,C/cm^2$ total charge has circulated; (c) final stage. $T = 920\,°C$; $Na_2SO_4 + 10\%\,NaCl$ molten salt mixture

The working electrode was always made the anode and operated at current densities ranging from 0 to $200\,mA/cm^2$ against a Pt/O_2 counter-electrode in a molten salt mixture of Na_2SO_4 and 10% NaCl at $920\,°C$. Electrode voltage was measured by an oxygen reference electrode (Pt/O_2 or stabilized zirconia).

Because the bath was saturated with respect to oxygen, the reference electrode potential can be considered to be unaffected by possible changes in bath composition during the measurements.

In Fig. 10 the results of potentiodynamic experiments on IN-738 LC samples with various coating treatments are summarized. The untreated samples were in the solution heat-treated condition, in order to avoid, as long as possible, the effects of γ-phase precipitates on corrosion rate.

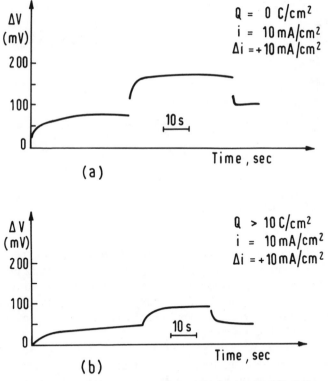

FIG. 13. Voltage vs. time plots for CeCr-aluminized IN-738 LC samples in galvanostatic conditions: (a) at the beginning of the run; (b) final stage. $T = 920\,°C$; $Na_2SO_4 + 10\%$ NaCl molten salt mixture

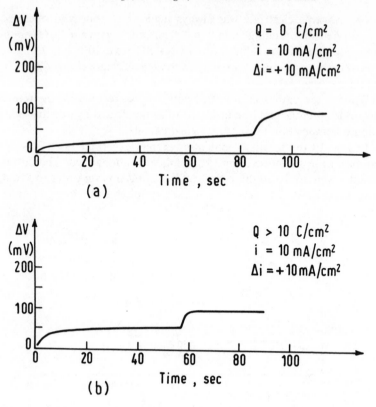

FIG. 14.   Voltage vs. time plots for Cr-aluminized IN-738 LC samples in galvanostatic conditions. Same as in Fig. 13

In no case was a true passivity condition observed and the sample behaviour is, generally speaking, 'active', but YCr-aluminizing processes generally led to eventual passivation, in contrast to the conventional Cr-aluminizing or to the Ce-modified Cr-aluminizing treatments. This may be related to the well-known effect of enhanced scale adhesion associated with Y additions to Ni-base superalloys.

Even more dramatic is the effect of the coating process on Hastelloy X which, in the uncoated condition, cannot withstand the attack of the molten salt mixture, even with no external voltage. When YCr-aluminized, it exhibits a corrosion behaviour similar to the best described above.

Measurements of current intensity in the potentiostatic mode (i.e. at

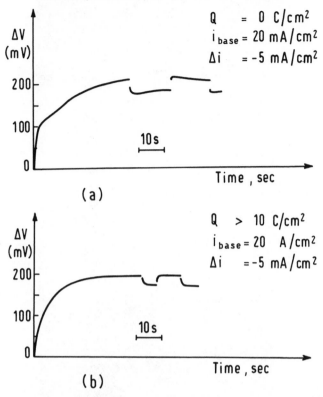

$Q = 0 \ \text{C/cm}^2$
$i_{base} = 20 \ \text{mA/cm}^2$
$\Delta i = -5 \ \text{mA/cm}^2$

10 s

Time , sec

(a)

$Q > 10 \ \text{C/cm}^2$
$i_{base} = 20 \ \text{A/cm}^2$
$\Delta i = -5 \ \text{mA/cm}^2$

10 s

Time , sec

(b)

FIG. 15.   Voltage vs. time plots for untreated IN-738 LC (solution treated) samples in galvanostatic conditions. Same as in Fig. 13

constant impressed overvoltage) provide a very simple and effective tool for evaluating corrosion rates in controlled but accelerated conditions.

In Fig. 11 such measurements are shown for protected IN-738 LC. On integration, a plot of the total circulated charge versus time, at constant overvoltage, is obtained: this is the electrochemical counterpart of a weight change versus time plot, but the advantage of directly obtaining the reaction rate versus time plot must be stressed.

All measurements have been performed starting from an 'active' condition to suppress the so-called incubation (or induction) time. This has been done because this time is dependent on the preformed oxide scale, the characteristics of which are very sensitive to accidental events.

In the galvanostatic mode, the electrode has been operated at controlled dc currents of various intensities, and the total electrode overvoltage has

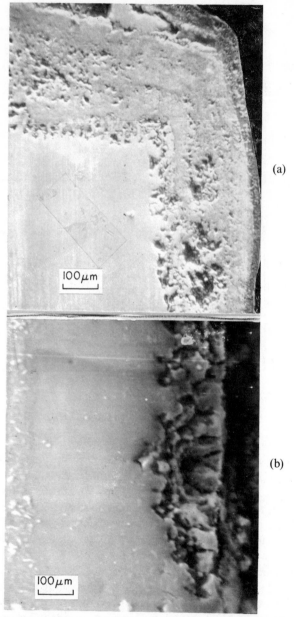

(a)

(b)

FIG. 16. (a) Duplex scale formed on a IN-738 alloy sample. $T = 920\,°C$; $Na_2SO_4 + 10\%$ NaCl molten salt mixture c.d. $5\,mA/cm^2$. 2 h. (b) Duplex scale formed on a YCr-aluminized IN-738 LC sample. Condition same as above

been measured in the conditions specified above. Application of current transients allowed the evaluation of the individual overvoltage components, conventionally specified as follows:

(1) Ohmic drop—relaxation time lower than instrument time constant.
(2) Ion exchange overvoltage—relaxation time a few milliseconds.
(3) Concentration polarization (c.p.)—relaxation time from dozens of milliseconds (in the bath salt) up to several minutes (in the scale).

Both oscillographic and fast (galvanometric) pen recordings have been obtained.

In Fig. 12 three diagrams summarize the voltage versus time characteristic and its variation with time for YCr-aluminized IN-738 LC.

The variations in electrical characteristics are as follows: at the beginning the ohmic component is very high and the c.p. is rapidly set up; at high circulated charges ohmic drop is negligible, while high c.p. overvoltages are slowly developed.

Peculiar behaviour is observed at intermediate circulated charges between the high and low levels: remarkable ohmic drops and presence of maxima in the polarization voltage. Electrochemists currently relate the last feature to the presence of two concurrent processes or to the nucleation of new structures. In the present case it can be tentatively related to the beginning of the growth of the duplex scale.

Both CrAl and CeCrAl coatings (Figs. 13 and 14) display such maxima very early after the current is switched on, but their behaviour is outstandingly better than untreated samples of IN-738 LC (Fig. 15).

Micrographic observation of corroded samples (Figs. 16a and b) substantiates the hypotheses and the conclusions advanced above.

DISCUSSION

H. Pflug (Robert Bosch GmbH, Stuttgart, Germany): Did you ever find Y and Ce as a metal in your scales or did the rare earths in the scales only consist of oxides? We know from the results of Seyboldt that the presence of rare-earth metals doesn't increase the corrosion resistance of Ni alloys.

Reply: No Y or Ce metals have been observed in the scale; only rare-earth metal oxidation products have been identified, mainly Y_2O_3.

In our opinion the presence of unoxidized rare-earth metals in the scale must be excluded on both thermodynamic and kinetic considerations, at least as long as the corrosion rate is not exceedingly high.

Please note, moreover, that the bromine–methanol solution, which has been used to separate the scale from the base metal, would certainly have dissolved any unreacted rare-earth metal present in the scale.

As to the second part of your question, the effectiveness of Y additions to Ni-base superalloys for improving their oxidation resistance is now being questioned by several authors.

In this case, however, we are dealing with impregnation coatings whose scaling characteristics are completely different from those of untreated Ni-base superalloys.

The resistance of fast thermal cycling corrosion to medium-term scaling in air and to sulphidation/oxidation was all enhanced by yttrium additions, although the rates in the very early stages of oxidation were increased.

19

Reaction-sintered Ni–Cr–Si Coatings on Nickel Alloys

E. FITZER, W. NOWAK and H.-J. MÄURER

Universität Karlsruhe, West Germany

ABSTRACT

The methods available for application of silicon-containing protective layers are discussed and a description is given of a new coating technique involving reaction sintering of overlay coatings. The influence of the composition of the overlayer and of the substrate alloy on the corrosion resistance of the layer has been investigated, using short-term thermobalance tests in air and long-term burner tests, both at 1000°C. Further studies of the effects of intermediate layers on the oxidation resistance of silicon-rich layers deposited on nickel-base alloys are also reported.

INTRODUCTION

For the protection of the iron metals and their alloys against high temperature oxidation, chromium, aluminium and silicon are suitable alloying elements, but up to now chromium and aluminium have been most widely applied for this purpose. These two elements can either be incorporated in the substrate, in an adequate amount, by alloying or the surface layer can be enriched. For the latter technique, aluminium and/or chromium coatings are used at present, depending on the field of application.

To protect iron against oxidation at temperatures of about 1000°C, contents of 20 at% chromium or 16 at% aluminium are necessary. Although protection against oxidation by silicon is obtained at much lower contents (~10 at%), silicon-containing coatings have not been applied practically up to now for several reasons. Silicon shows a very much higher reactivity to iron. As a consequence of this, asymmetric interdiffusion of

313

314   *E. Fitzer, W. Nowak and H.-J. Mäurer*

both elements occurs leading to immediate impairment of the coatings by the Kirkendall effect. Later low melting Fe–Si–O eutectics are formed during oxidation (1–3).

These Kirkendall effects can be suppressed by increasing the Ni content in the substrate. As a result, silicide coatings on nickel alloys have pore-free structures (4) but nevertheless these layers are very brittle. In unfavourable coating conditions low melting Ni–Si eutectics will also occur.

In spite of these limitations, the application of silicide coatings is being considered more and more. This is due to the resistance of silicon-containing coatings to high temperature corrosion generally (5) and especially to V_2O_5 corrosion. This resistance seems to apply to silica-forming coatings as well as to those forming Ni_2SiO_4 or Fe_2SiO_4 (5–7).

PREPARATION METHODS FOR SILICON-CONTAINING COATINGS

Figure 1 shows the different possibilities for preparing silicon-containing coatings. In principle we can distinguish between diffusion layers and overlay coatings. In the case of diffusion methods like pack cementation or some special chemical vapour deposition (CVD) processes, the layer growth occurs within the original substrate surface. In some of our experiments the preferential formation of silicides with low chromium contents, caused by reaction between silicon and the substrate, was observed but our main

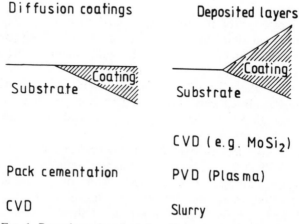

FIG. 1. Procedures for obtaining silicon-containing coatings

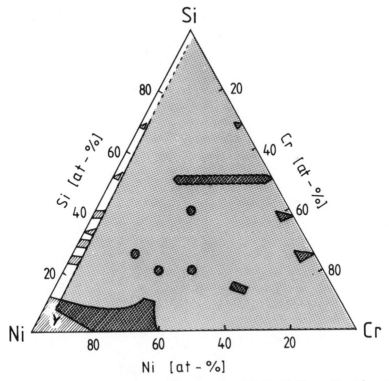

FIG. 2. Phase diagram of the ternary system Ni–Cr–Si (from Knotek and Lugscheider, 1974). The field giving adequate resistance to oxidation is distinguished by the shaded area

interest was focused on overlay coatings whose performance should be independent of the composition of the base material. In the case of these coatings, however, problems of chemical and mechanical compatibility with the substrates can arise. $MoSi_2$ deposited in a CVD process is not suitable for the protection of nickel alloys, because of its low thermal expansion as well as its high reactivity with the substrate. But the same material can be applied successfully to niobium or tantalum substrates (8, 9).

In our experience the most suitable silicon-containing coatings for nickel alloys were Ni–Cr–Si layers deposited by plasma spraying or by a slurry process. Therefore, in our experiments we measured the growth rates of oxide layers on Ni–Cr–Si alloys, confining our attention to oxidation in air containing no other high temperature corroding species.

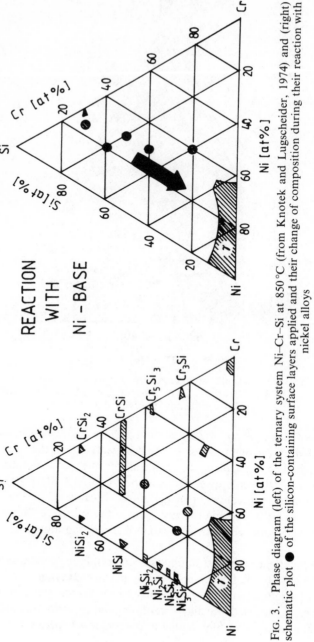

FIG. 3. Phase diagram (left) of the ternary system Ni–Cr–Si at 850 °C (from Knotek and Lugscheider, 1974) and (right) schematic plot ● of the silicon-containing surface layers applied and their change of composition during their reaction with nickel alloys

It was shown that a concentration of 8–10 at% silicon in the γ-solid solution with a small amount of chromium will guarantee sufficient protection against oxidation. Moreover, the Cr content of 20 at%, necessary for good protection of silicon-free nickel alloys, can be fully substituted by half this concentration of silicon, i.e. 10 at%.

The shaded part of Fig. 2 indicates the field of the ternary Ni–Cr–Si system which is supposed to guarantee adequate resistance to high temperature oxidation. In this field the resistance of Ni–Cr–Si alloys increases with increasing Si and Cr contents. On the other hand, pure Ni silicides with little or no chromium (white area of Fig. 2) show insufficient oxidation resistance.

The oxidation studies proved that silicide coatings on nickel alloys should be prepared from Ni–Cr–Si phases in equilibrium with the γ-solid solution. On the other hand, phases with higher silicon and chromium contents can usefully be applied to compensate for the effect of dilution by nickel during the subsequent heat treatment. Another experimental study was undertaken to show whether the coating compositions under consideration were compatible with the substrates and which manufacturing process would be the most suitable one.

COATINGS OF ALLOYED Ni–Cr–Si POWDERS

The right-hand phase diagram in Fig. 3 shows the compositions and compounds applied by plasma spraying in an argon atmosphere. The 100–200 μm thick layers (left-hand side of Fig. 4) did not adhere to the substrate during oxidation. However, during heat treatment in hydrogen at 1000 °C a very violent reaction took place accompanied by a partial melting process at the surface of the specimen (right-hand side of Fig. 4).

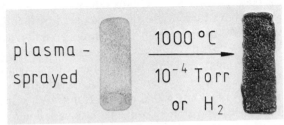

FIG. 4. Surface layer of high silicon content applied to Nimonic 90 by powder plasma spraying. Left: after plasma spraying; right, after the melting reaction during the heat treatment

The changes of composition occurring in the silicon-containing coatings during the reaction with the nickel alloys are indicated schematically by the black arrow in the right-hand phase diagram of Fig. 3. Almost independently of the composition of the various coating alloys, the reaction leads to very thick diffusion zones (600–800 μm), with similar compositions to the γ-solid solution, with isolated precipitates of high chromium and

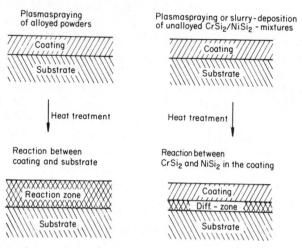

FIG. 5. Procedures suitable for applying surface layers with high silicon and chromium contents

silicon contents. Although these layers exhibit high oxidation resistance, they must be regarded as unfavourable because of their very high penetration depth and resulting embrittlement of the substrate.

The melting reaction between the substrate surface and the alloyed coatings described above can only be slightly reduced by modification of the process parameters. In order to obtain sufficient adhesion between the coating layer and the substrate it is necessary to achieve a temperature of at least 1000 °C for starting the reaction between the coating components Cr and Si and the nickel from the substrate. As a result, the reaction temperature increases to about 1450 °C in a short time. When the reaction has started, it is nearly impossible to control it by a variation of treatment time and temperature. The whole coating process of plasma spraying of alloyed powders, followed by the reaction during the heat treatment, is shown schematically on the left-hand side of Fig. 5.

REACTION SINTERING OF THE COATINGS

In order to avoid the problems with the melting reaction, the procedure shown on the right-hand side of Fig. 5 has been chosen for the following experiments. An attempt was made to restrict the reaction to the surface layer, thus avoiding damage to the substrate. The heat of the reactions is then utilized to sinter the outer layer to a dense coating. This is only possible if unalloyed powder mixtures are applied to the substrate.

To get high chromium and silicon contents in the coating layers we used $CrSi_2/NiSi_2$ mixtures of various compositions. As expected, isostatic pressed mixtures of $CrSi_2$ and $NiSi_2$, heat treated at 1000 °C (hydrogen or argon), formed compacts with low porosity.

These powder mixtures were applied to the base materials by a slurry process because in plasma spraying the possibility of a premature reaction could not be precluded. The powder was suspended in an organic binder and painted on the specimens; the latter were annealed *in vacuo*. Figure 6 shows cross-sections of such reaction-sintered layers on two different nickel alloys. With Nimonic 90 as the base material, damage of the substrate could not be prevented. But in comparison with samples coated by alloyed plasma-sprayed layers the effect of the reaction was obviously reduced. The reactivity of the $CrSi_2/NiSi_2$ mixtures can also be decreased by an increase of the $NiSi_2$ content.

By using other substrate alloys it was proved that the melting reaction also depends on the composition of the substrate. Furthermore, during these experiments it was shown that some elements present in the base alloys were able to reduce the melting reaction.

Alloys with high aluminium contents exhibit good chemical compatibility with $CrSi_2/NiSi_2$ layers. Within certain limitations this applies to the alloy IN-738 with about 3.5 wt % Al. In the case of IN-100 with a very high γ'-content due to the 6 wt % Al, the damage to the substrate has been completely suppressed (Fig. 6a). Obviously, the associated formation of a NiAl layer is responsible for this behaviour (see lower part of the cross-section). A very similar effect could be found in the case of alloys containing molybdenum and tungsten.

Possible methods for applying silicon-containing coatings to substrates with low contents of these reaction-reducing elements (e.g. Nimonic 90) are shown schematically in Fig. 7.

In Fig. 7 (left) an unalloyed powder mixture of a composition corresponding to the γ-solid solution reacts with the substrate surface. Such solid solution layers should not undergo a damaging melting reaction with

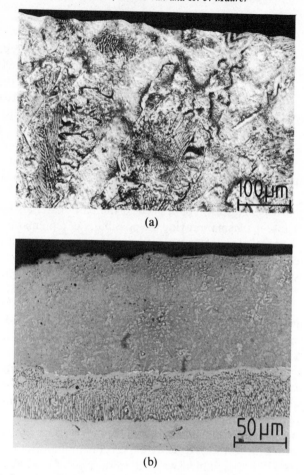

(a)

(b)

FIG. 6. Influence of the composition of the substrate on the chemical compatibility between high silicon surface layers and substrate: (a) Nimonic 90 (1.4% Al); (b) IN-100 (5.6% Al)

any base material and have the advantage of good ductility. Their disadvantage is their reduced long-term oxidation resistance due to their low silicon content.

Figure 7 (right) shows a second approach designed to give high chromium and silicon contents in the surface layer. This is done by controlling the interdiffusion of the silicon or chromium from the layer and the nickel from the substrate by applying an additional barrier layer

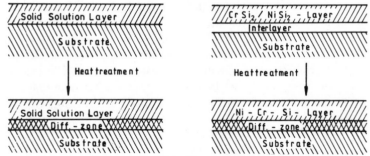

FIG. 7. Procedures for applying high silicon coatings to alloys without elements weakening the reaction (e.g. Nimonic 90)

between the substrate and the coating to weaken the violent reaction. These layers have a high silicon content to provide good long-term oxidation resistance. The only disadvantage is their relatively high brittleness.

In the following section both possibilities for a solution to this problem are discussed in detail.

SOLID SOLUTION LAYERS WITH LOW SILICON CONTENT

The preparation of low silicon solid-solution layers was accomplished by reaction sintering of coatings of $Ni/Cr/CrSi_2$ mixtures. While compacts made of $Ni/20Cr/10Si$ or $Ni/30Cr/10Si$, respectively, showed a small residual porosity, the layers formed from these powder mixtures were free of pores (Fig. 8). The substrate had not been damaged. In the solid solution

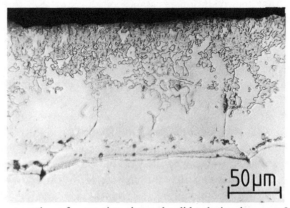

FIG. 8. Cross-section of a reaction-sintered solid-solution layer on Nimonic 90

layers, silicide precipitates with high Cr content had been formed. This is a
similar layer structure to that found in the case of the melting reaction
between plasma-sprayed alloys and Nimonic 90. The advantage of the
reaction-sintered slurry layer arises because the melting reaction takes place
inside the layer without damaging the substrate surface. The oxidation
resistance of these samples at 1000 °C was much better than that of the
compacts and this is obviously due to the porosity of the compacts. It was

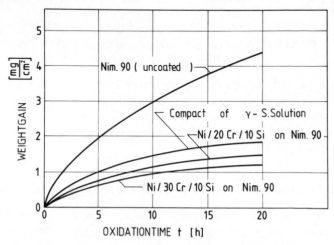

FIG. 9.   Isothermal short-term oxidation tests (1000 °C, flowing air) of compacts
and coatings made of γ-solid solution

also shown that the oxidation resistance of coatings with similar silicon
contents increases with higher chromium contents (Fig. 9).

One disadvantage of these solid-solution layers is that they only provide
a limited reservoir of silicon, so an attempt was made to incorporate high
melting silicides like $TiSi_2$ into the coating. This is quite easy to realize up to
levels of 10 wt % $TiSi_2$. Similar coating systems are commercially available
as Nicrocoat 620 and 621.

The layer structure and the distribution of the elements in a briefly
oxidized (1100 °C) Nicrocoat 621 layer on IN-601 are shown in Fig. 10 as
scanning electron micrographs using emitted secondary electrons (SEI) and
elemental scans using the specific X-rays (11). During the oxidation the
precipitates in the coating do not change significantly. The $TiSi_2$ is
distributed in the form of precipitates all over the layer and is mostly located
in small cavities. During a longer heat treatment it decomposes partially,

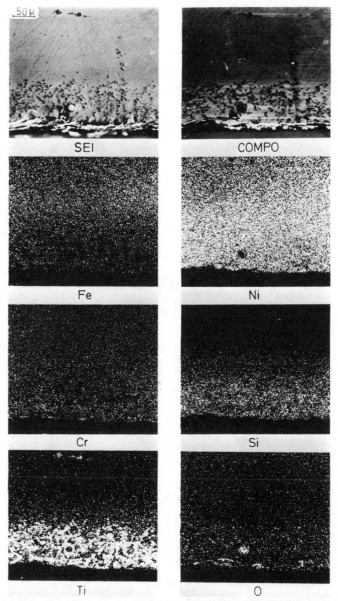

FIG. 10. Nicrocoat 621 layer on IN-601 (oxidized 50 h at 1100 °C). Scanning images obtained by the secondary electrons (SEI), a combination of secondary and back-scattered electrons (COMPO) and the specific X-rays of the elements (Ni, Fe, Cr, Si, Ti, O)

the silicon going into the surrounding matrix. Oxidation tests at 1100 °C lead to internal oxidation followed by the formation of titanium oxides which are not protective against oxidation at these temperatures. An increase of the TiSi$_2$ content beyond 10 wt % must be avoided because the sintering behaviour gets worse. In oxidation tests the coatings become detached after a short time.

COATINGS WITH HIGH SILICON CONTENTS AND ADDITIONAL REACTION BARRIERS

In this section the application of high silicon coatings on Nimonic 90 with an additional reaction barrier interlayer will be discussed. For this purpose chromium enrichment of the substrate's surface, produced either by electrolysis or pack cementation proved to be useful.

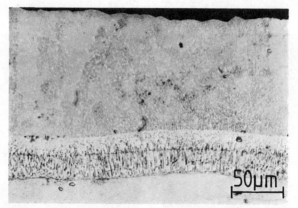

Fig. 11.   Reaction-sintered layer with high silicon and chromium content on Nimonic 90 with Cr interlayer

Figure 11 shows such a reaction-sintered layer. Although the chromium layer has been completely consumed during the sintering reaction, the substrate is not detectably damaged. The outer layer with high silicon content is separated from the diffusion zone by a silicide phase of high chromium content, which acts as a diffusion barrier. On such a chromized substrate the coating composition can be varied to a large extent, between pure NiSi$_2$ and a mixture of 70 wt % CrSi$_2$/30 wt % NiSi$_2$ without changing the structure.

CrSi$_2$ contents of more than 70 wt % impede the sintering so that the outer layer becomes more and more porous. Above 90 wt % CrSi$_2$ no more sintering takes place in the outer part of this layer. Also, aluminium is a reaction-weakening element for the interlayer. The melting reaction, discussed above, of a mixture of 70 wt % CrSi$_2$/30 wt % NiSi$_2$ on IN-100 allowed the formation of a NiAl interlayer, so this implies that aluminized Nimonic 90 would behave in the same way. Furthermore, aluminium additions have the beneficial effect of introducing another oxidation-resistant element into the coating.

FIG. 12. Reaction-sintered layer with high silicon and chromium content on aluminized Nimonic 90

Figure 12 shows such a specimen which has been coated with a slurry of 70 wt % CrSi$_2$/30 wt % NiSi$_2$ after aluminizing. The structure of the sample resembles that of a specimen which has only been aluminized. In the inner part of the coating the typical chromium enrichment can be observed (light precipitates) followed by a small zone of NiAl, which is free of silicon. The major part of the layer (dark area) consists of silicon-containing NiAl (4–8 wt % Si) with high silicon and high chromium concentrations in the precipitates (light). On the surface, a thin layer of a CrSi phase could be detected by microprobe analysis. Only a thin CrSi$_2$/NiSi$_2$ layer had been applied to the specimen shown in Fig. 12.

For thicker coatings, manufactured under certain conditions, the whole aluminide layer may be consumed during the sinter reaction. The aluminide and silicide phases segregate resulting in globular precipitation of NiAl in a silicide matrix. But even in such a case the substrate surface is not noticeably damaged. Chromium-aluminized layers as interlayers, instead of pure aluminized ones, lead to the same result.

OXIDATION TESTS

In oxidation tests it was proved once again that high silicon contents alone are not sufficient to give high oxidation resistance, but that high chromium contents are also necessary. For example, during oxidation for 20 h at 1000 °C chromized Nimonic 90 coated with pure $NiSi_2$ showed a higher weight gain than a specimen coated with a slurry of $CrSi_2/NiSi_2$, the silicon content being approximately the same in both cases (Fig. 13). Furthermore, it was shown that in similar short-term oxidation tests samples with aluminium interlayers had better resistance than those with chromium interlayers. The composition of the slurry applied was exactly the same. However, both coating systems proved to be more oxidation resistant than the γ-solid-solution layer (Fig. 14).

Figure 15 shows the long-term oxidation behaviour of the various coating systems in burner gas (natural gas with excess air) at temperatures between 960 and 1000 °C. Several substrates coated with solid-solution layers, as well as $CrSi_2/NiSi_2$ with additional interlayers, were tested. To get comparable results a coating with constant composition ($50CrSi_2/50NiSi_2$)

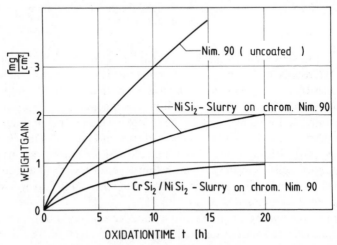

FIG. 13.   Isothermal short-term oxidation tests (1000 °C, flowing air) of high silicon-containing layers on Nimonic 90 with Cr interlayer

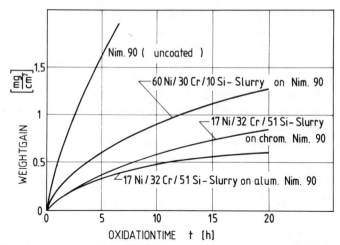

FIG. 14. Comparison of the various coating systems in short-term oxidation tests at 1000 °C (flowing air)

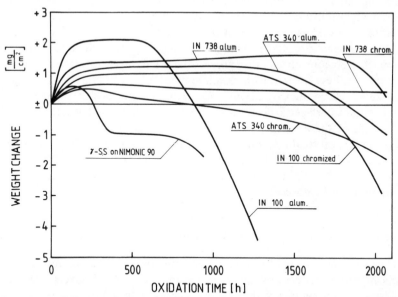

FIG. 15. Long-term oxidation behaviour of the various coating systems at 1000 °C in burner gas (natural gas with excess air). All samples, except the one with a γ-solid solution layer, had been coated with a surface layer of 26 wt % Ni, 24 wt % Cr, 50 wt % Si (= 50 wt/% CrSi$_2$/50 wt % NiSi$_2$). ATS 340 is equivalent to Nimonic 90

was applied. Electrolytically deposited chromium and CrAl layers made by pack cementation were used as interlayers. As was to be expected, the coatings with high silicon contents showed longer lifetimes than the lower silicon γ-solid-solution coatings. At present, it cannot be definitely decided which of the layers tested will have the best oxidation resistance and longest lifetimes. In the case of Nimonic 90 (equivalent to ATS 340) both coating systems exhibited similar behaviour. For substrates with high aluminium content the coating system with chromium interlayers seemed to be superior to that with CrAl interlayers. This applies especially for IN-100.

DIFFUSION BEHAVIOUR AND ITS INFLUENCE ON THE MECHANICAL PROPERTIES OF THE WHOLE COMPOSITE

When annealing samples with overlay coatings consisting of $50CrSi_2/50NiSi_2$, it was observed that the previously deposited interlayers did not act as true diffusion barriers. In both cases (Cr interlayer, CrAl interlayer) the diffusion of silicon into the substrate is relatively fast (about 200 μm after 100 h at 1000 °C). In these cases the interlayers will be completely consumed. The inward diffusion of the silicon, combined with its reaction with the elements of the interlayer, leads to a decrease of the room temperature (RT) flexural strength of the whole composite. Immediately after the coating process the RT flexural strength of samples (total thickness 1.5 mm) with aluminium-containing interlayers was found to have deteriorated by about 50%, while in the case of samples with chromium interlayers a decrease of only 10% was observed. But the flexural strength of these samples with chromium interlayers is also decreased further by heat treatment due to the inward diffusion of silicon. Only in the case of the low silicon solid-solution layers could such a major influence of the coating process and the subsequent heat treatment on the RT flexural strength not be observed.

SUMMARY

The usual application of alloyed coating materials to metallic substrates by plasma spraying does not seem to be feasible for Ni–Cr–Si layers with high silicon content on nickel alloys. This is because during the necessary heat treatment either no sintering takes place or the substrate is injured by a violent reaction between layer and base material.

By the use of unalloyed or partially alloyed silicide powder mixtures it is possible to keep the reaction inside the layer which thus forms a nearly non-porous coating. By application of powders with a composition corresponding to the γ-solid solution, layers are obtained which are compatible with all metallic substrates. These low silicon coatings are not oxidation resistant in long-term exposure at 1000 °C, but for some applications they can be good enough at lower temperatures.

Samples coated with non-porous layers with high silicon and chromium contents show good long-term oxidation resistance. Although the sintering reaction proceeds mainly inside the layer, a certain amount of reaction with nickel alloys is still observed. This is the reason why these layers are only applicable to those materials which have special constituents able to form reaction barriers spontaneously (e.g. the aluminium, tungsten and molybdenum in IN-100 and MAR M002).

The other alloys, such as Nimonic 90, must be covered with reaction barriers as an additional step before coating with layers of high silicon content. Either chromized or aluminized or chromium-aluminized layers proved to be suitable. The layers with high silicon content are relatively brittle and diffusion of Si from them into the substrate will cause embrittlement of the latter, especially if Al is also present. Apparently, the interlayers tested up to now do not operate as diffusion barriers and therefore it is necessary to look for new interlayer systems.

REFERENCES

1. E. Fitzer and M. Niessner (1951). *Metallk. Ber.*, **23**, 3–35.
2. E. Fitzer (1954). *Arch. Eisenhütt Wes.*, **25**(9/10), 455–63.
3. E. Fitzer (1953). *Z. Metallk.*, **44**, 462–72.
4. E. Fitzer (1955). *Arch. Eisenhütt Wes.*, **26**, 159–69.
5. E. Fitzer and J. Schwab (1953). *Berg-u. Hüttenm. Jb.*, **98**, 1–7.
6. J. Schlichting (1975). *Z. Werkstoff.*, **6**(1), 11–16.
7. P. Felix and E. Erdoes (1972). *Werkstoffe Korros.*, **23**(8), 627–36.
8. D. Kehr (1975). Diss., Inst. for Chem. Techn., University of Karlsruhe.
9. K. H. Kochling (1976). Diss., Inst. for Chem. Techn. University of Karlsruhe.
10. O. Knotek and E. Lugscheider (1974). Preprints 8th Plansee Seminar, p. 20.
11. E. Fitzer, H.-J. Mäurer and J. Schlichting (1975). *Z. Metallk.*, **66**(10), 565–9.

DISCUSSION

K. W. Wegner (University of Münster, Germany): Have you tested the thermal shock behaviour of your layers? Were the oxidation tests carried out continuously, that is isothermally?

Reply: In the case of the short-term oxidation tests, the samples were tested continuously, that is isothermally in a thermobalance, in the usual way. The heating rate up to the final temperature of 1000 °C was 25 °C/min.

In the case of the long-term oxidation tests the samples were exposed at 1000 °C to burner gas. In these tests the specimens were taken out of the burner every 24 h and cooled down in air. Their weight change was measured using an analytical balance. After that the samples were put back into the hot burner gas. Even with these abrupt changes of temperature no tendency for spalling of the layers was found with most of the samples tested.

Only in the case of aluminized IN-100 was a relatively high weight loss observed, which was obviously due to spalling of material from the layer.

R. Pichoir (ONERA, F-92320, Chatillon, France): (1) Is your investigation of the siliconizing of IN-100 of a fundamental nature? If it is not, what is the envisaged industrial application? (2) In your particular case what is the exact significance of 'diffusion barrier layer'?

Reply: (1) Our investigation of the siliconizing of IN-100 is a fundamental study. We are only interested in the influence of the composition of the base alloy on the diffusion rate of silicon in such alloys. However, because of the good resistance of our Ni–Cr–Si layers to oxidation, there may be a possible application to aircraft turbines in the future.

(2) The interlayers mentioned in our paper will act only as reaction barriers. That means they are intended to suppress the violent reaction between coatings with high silicon and chromium contents and the substrate during the sintering reaction of $CrSi_2$ with $NiSi_2$ in the surface layer. This sintering reaction is accompanied by a large evolution of heat which will start the reaction between coating and substrate. These reaction barriers do not decrease interdiffusion sufficiently and therefore it is necessary to find new diffusion barriers.

A. R. Nicoll (BBC, Heidelberg, Germany): You reported that these two layer systems are brittle. How did you determine this?

Reply: In previous studies we have tested the RT flexural strength of compacts, whose composition corresponded to the surface layers later used. In that work the samples with high Cr and Si contents showed a marked brittle fracture behaviour, while samples with low Si contents having the γ-solid-solution structure had better ductilities. These results were confirmed by similar experiments with coated samples of Nimonic 90.

The layers with high Cr and Si contents showed a high tendency to crack formation in the tensile zone, but this was not observed for the solid-solution layers even after severe bending.

We have not yet determined the temperatures at which the brittle layers will show a transition to ductile behaviour. The strength of the total system will be reduced much more by layers with high Cr and Si contents than by γ-solid-solution layers. The samples with aluminide interlayers showed a higher decrease of strength than those with Cr interlayers.

20

Preparation and Investigation of Layers Enriched in Silicon by Chemical Vapor Deposition

G. WAHL and B. FÜRST

Brown Boveri & Cie AG, Heidelberg, West Germany

ABSTRACT

Siliconizing by the reaction of $SiCl_4$ or $SiHCl_3$ with metal surfaces is an important process for the preparation of silicon-enriched layers. The influence of the process parameters (temperature $900°C < T < 1200°C$, chemical vapor deposition (CVD), with or without electrical discharge) on the deposition process (deposition kinetics, liquid and solid phase formation, phase composition) is studied. The comparison of the experimental results with thermodynamic calculations shows that the deposition is controlled mainly by chemical reactions which take place on the surface.

Parts of this chapter were presented at the 6th Conference on Chemical Vapor Deposition (Atlanta, 1977) and have been published in the Proc. 6th CVD Conference (*ECS, Princeton, 1977*).

INTRODUCTION

Siliconized layers on alloys are an interesting subject for investigation because of their excellent corrosion properties (1–5). The preparation of layers by CVD (1,4,6) is described in this chapter; in the process investigated the silicon is transported in the form of chlorides ($SiCl_4$, $SiHCl_3$) through the gas phase (H_2) on to the metal surface (e.g. Ni, Co, Nim 105, IN-738). The investigation of the siliconizing process concentrates on two points:

(1) The thermodynamic analysis.
(2) The kinetic problems of the siliconizing process; in this case the mass transfer of Si from the gas phase on to the metal is the main point of interest.

THERMODYNAMIC ANALYSIS

Thermodynamic calculations for the siliconizing process are very instructive because they give information on (a) the yield (ε) of the Si deposition from the gas phase at a given gas composition, and (b) the competition between the siliconizing process and metal evaporation enhanced by the formation of gaseous metal chlorides.

For these calculations the H_2–$SiCl_4$ gas mixture (mole per cent of $SiCl_4 = x_0(SiCl_4)$) is assumed to be equilibrated above the heated metal

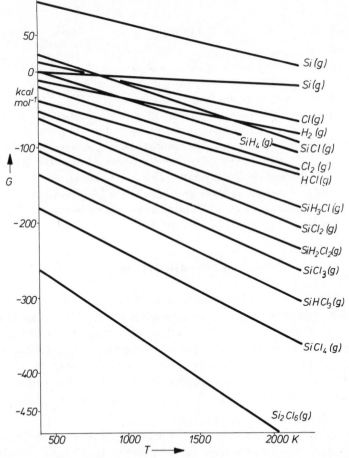

FIG. 1.   Free enthalpies of SiHCl compounds (7, 8): g, gaseous; c, condensed

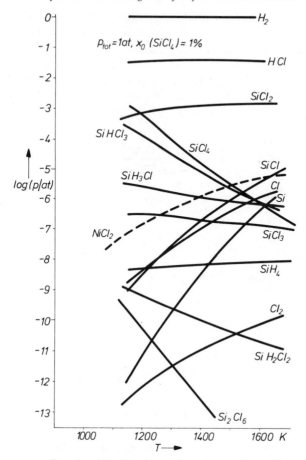

FIG. 2. Calculated partial pressures above Ni

surface which was taken to be Ni, Cr or Fe. Both gas and metal may have the same temperature, T. In thermodynamic equilibrium a number of gaseous (g) and condensed (c) compounds and atoms are formed: $Si(g)$, $Si(c)$, $Cl(g)$, $Cl_2(g)$, $H_2(g)$, $SiCl(g)$, $SiCl_2(g)$, $HCl(g)$, $SiCl_3(g)$, $SiH_3Cl(g)$, $SiH_2Cl_2(g)$, $SiHCl_3(g)$, $SiCl_4(g)$, $Si_2Cl_6(g)$.

In addition, metal chlorides exist. In the case of Ni: $NiCl_2(g)$; in the case of Cr: $CrCl_2(g)$, $CrCl_3(g)$; and in the case of Fe: $FeCl(g)$, $FeCl_2(g)$, $FeCl_3(g)$ and $Fe_2Cl_6(g)$. The formation of condensed chlorides can be neglected because of the very high vapor pressure of these chlorides. The atomic evaporation of Cr, Ni and Fe is neglected because their influence is small in the range of the siliconizing temperatures. In these calculations it is assumed that silicon deposition takes place on the metal surface and that the formation of silicides can be neglected.

The free enthalpies (shown in Fig. 1) of the species listed are taken from ref. 7 for the SiHCl compounds and for all other compounds from ref. 8. For the computation of the partial pressures and the fraction of the condensed phases a modified version of a computer program described in ref. 9

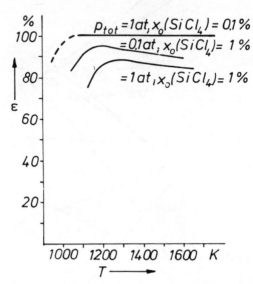

FIG. 3.   Yield ε vs. temperature T, calculated for Si deposition on Ni

was used. This program minimizes the free enthalpy of the system through an iterative technique, including gaseous and condensed phase species. A condensed phase species is automatically excluded from the computational process, when its concentration has reached a preset lower limit.

Figure 2 shows the temperature dependence of the partial pressures of gaseous components formed above a Ni surface at $x_0(SiCl_4) = 1\%$ and the total pressure $p_{tot} = 1$ atm. The prevailing gaseous Si compounds are $SiHCl_3$, $SiCl_4$, $SiCl_2$ at low temperatures and only $SiCl_2$ at high temperatures. In between there is a maximum yield ε shown in Fig. 3; the yield ε is defined as the amount of Si which is deposited divided by the whole silicon content in the gas phase before the equilibration. The yield ε is always $>80\%$ and depends only weakly on $x_0(SiCl_4)$ and p_{tot}. If the formation of silicides had been included in the calculations the yield would

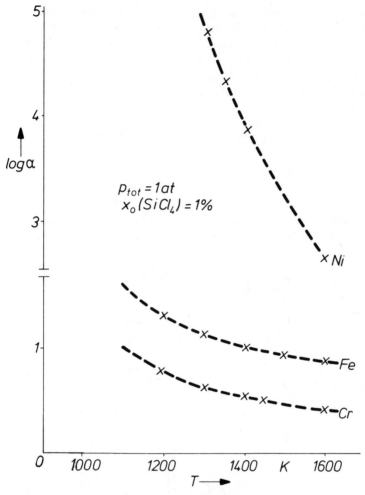

FIG. 4.   Ratio α vs. temperature T calculated for Si deposition on Ni, Cr and Fe

be still higher, because of the negative free enthalpies of silicide formation (10). The competition between siliconizing and metal evaporation is described by the quantity α which is defined as the ratio of the depositing silicon atoms and the evaporating metal, as shown in Fig. 4. The factor α was calculated for the siliconizing of Ni, Cr and Fe. The evaporation can be neglected in the case of Ni but not in the case of Fe and Cr.

To sum up, the calculations show that silicon deposition and therefore siliconizing is possible at temperatures $T > 1000\,K$ when kinetic restrictions are excluded. The task of the experimental part of this chapter is to investigate the kinetics of the siliconizing process. The competition of evaporation of metal and siliconizing is of importance for siliconizing superalloys, because these alloys contain large amounts of Cr. This will be discussed in the experimental part.

EXPERIMENTAL ARRANGEMENT

A schematic diagram of the deposition arrangement is shown in Fig. 5. The hydrogen cleaned in a palladium cell was led through a $SiCl_4$ bottle where the gas was enriched with $SiCl_4$. The $SiCl_4$ flow could be determined continuously by measuring the $SiCl_4$ level in the bottle. For the $SiHCl_3$ siliconizing the $SiHCl_3$ flow was measured separately with a thermal gas

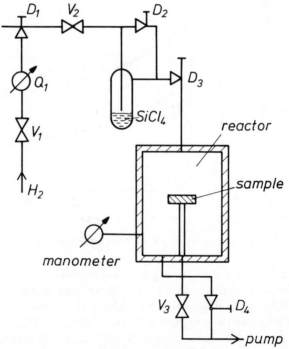

Fig. 5.   Schematic diagram of the siliconizing apparatus: D, throttle valve, V, valve, Q, flowmeter

flowmeter. The pressure in all reactions could be varied between 10^{-3} torr and 760 torr. For the siliconizing three reactions were used:

(1) A quartz reactor ('hot wall reactor'), which is heated in a resistance furnace (50 cm long, 6 cm diameter).

(2) A 'cold wall reactor', in which only the sample was heated to the deposition temperature T_d by radio frequency (10 kc). The walls of the reactor were cooled by water.

(3) A reactor, in which the sample was heated in an anomalous glow discharge (11). The sample was the cathode, which was heated by bombardment of ions accelerated in the cathode drop region. In this case the chloride–H_2 mixture is highly activated by the discharge in this gas.

EXPERIMENTAL RESULTS AND DIFFUSION CALCULATIONS

The main quantity measured was the mass deposition rate $\dot{m}$ of Si determined by the expression:

$$\dot{m} = \frac{m_a - m_b}{t} \qquad (1)$$

where m_a is the mass of the sample after, and m_b the mass of the sample before, the siliconizing process, and t is the siliconizing time (mostly $t = 2$ h). The Si transport is correctly described by eqn. (1) provided that there is no transport of the sample material into the gas phase. This condition applies for the case of Ni, as shown in Fig. 4. The same applies for Co because of the similar thermodynamic properties of Ni and Co. For Fe- and Cr-containing alloys the equation is only approximately correct because of the transfer of Cr and Fe into the gas phase.

Figure 6 shows siliconizing experiments on Ni. The mass deposition rates are displayed against the reciprocal temperature T_d for the siliconizing processes performed in the different reactors. The deposition conditions were:

Total gas flow: 33 l/h (STP), $x_0(SiCl_4) = 1\%$.
Total pressure in the hot and cold wall reactor: 200 torr.
Total pressure in the gas discharge reactor: 20 torr.
Geometry of the sample: 20 mm diameter, 5 mm thick.

As shown in Fig. 6 the surface of the sample was melted above a characteristic temperature T_c during the deposition process. A typical

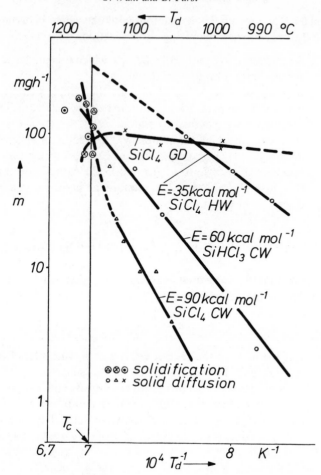

FIG. 6.   Temperature dependence of the Si deposition rate $\dot{m}$: GD, glow discharge; CW, cold wall reactor; HW, hot wall reactor; $x_0(SiCl_4) = x_0(SiHCl_3) = 1\%$; $i_{tot} = 33\,l/h$ (STP); CW, HW, $p_{tot} = 200\,torr$; GD, $p_{tot} = 20\,torr$

metallographic section of the solidification structures formed is shown in Fig. 7. Electron microprobe analyses and the good correlation with the phase diagram reveal that a Ni–Ni$_3$Si eutectic structure is formed. The value T_c given in Fig. 6 is identical with the eutectic temperature T_{eut} (Ni–Ni$_3$Si).

At temperatures $T_d < T_c$, however, the silicon deposited diffuses by solid-state diffusion into the Ni sample and silicides are formed (Fig. 8).

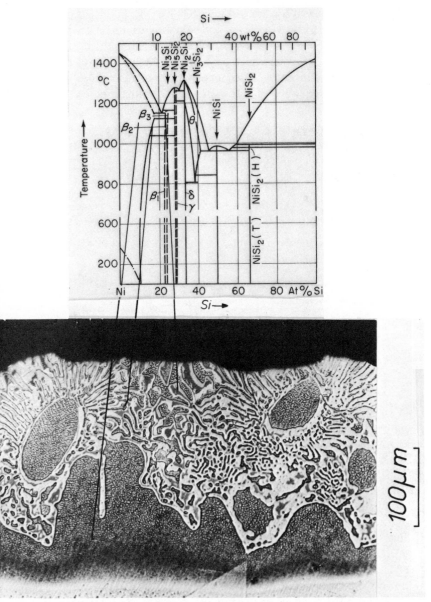

FIG. 7. Correlation of the Ni–Ni₃Si solidification structure on a Ni sample with the phase diagram (12). The large grains of Ni in the eutectic structure are remains of the original Ni structure

Depending on the silicon content ($x_0(\text{SiCl}_4) < 3\%$) in the gas phase, Ni_3Si and Ni_5Si_2 were formed in the hot and cold wall reactors. Compounds with a higher content of Si (Ni_2Si, Ni_3Si_2, NiSi, NiSi_2, Si) were only formed with glow discharge siliconizing. These phases were identified by X-ray diffraction. The deposition in the hot and cold wall reactor can be described by an Arrhenius equation $\dot{m} = k_0 \exp(-E/RT)$ with the activation energies

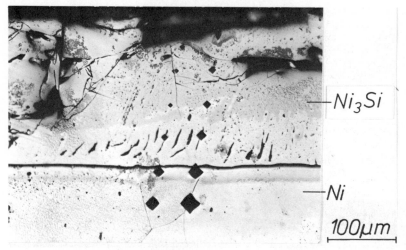

FIG. 8.   Silicide layer formed on the Ni surface by solid-state diffusion

E given in Fig. 6. Only with glow discharge siliconizing was a very weak temperature dependence found.

The siliconizing process is not only dependent on the reactor but also on the Si–chloride compound used. This is shown in Fig. 6, in which the results for siliconizing with a SiHCl_3–H_2 mixture are displayed alongside those for siliconizing with a SiCl_4–H_2 mixture. In spite of the equal deposition conditions SiHCl_3 siliconizes at lower temperatures and shows a smaller activation energy, $E = 60\,\text{kcal mol}^{-1}$. A similar effect has been found by Eversteyn (13), i.e. an increase in the deposition of Si if Si chlorides with a decreasing number of Cl atoms were used (SiCl_4, SiHCl_3, SiH_2Cl_2, SiH_4).

The critical temperature T_c, above which the metal surface is melted during the siliconizing process, is $T_c = T_{\text{eut}}$ (Ni–Ni_3Si) only at values $x_0(\text{SiCl}_4) > 0.5\%$, as shown in Fig. 9. At smaller $x_0(\text{SiCl}_4)$ the surface is not melted because the deposited silicon is in solution in the Ni. The boundary between the solid-solution range and the melting range, which is shown in

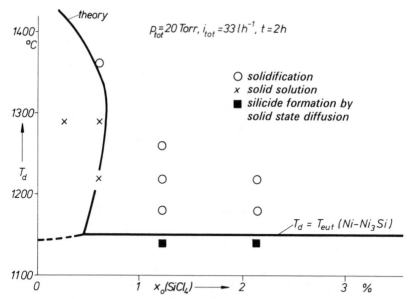

FIG. 9. Boundary between the different siliconizing ranges. The experiments were performed by glow discharge siliconizing. The boundary marked 'theory' was calculated

Fig. 9, is calculated by the numerical solution of the one-dimensional differential equation for diffusion (see eqn. (5) later).

The dependence of the siliconizing process on the substrate material is shown in Fig. 10, in which the mass deposition rate is plotted against the temperature for Ni, Co, Mo and the superalloy Nim 105 (20Co/15Cr/5Mo/1Fe/5Al/1.0Ti/1.0Si, Ni balance, wt%). The eutectic temperatures of Ni and Co are also given; in the case of Mo all eutectic temperatures ($T > 1400\,°C$) are out of the temperature range shown in Fig. 10. The phase diagram of Nim 105–Si is unknown. The curves of all materials resemble one another and only the position on the temperature scale is different. A comparison of the siliconizing curves in Fig. 10 with the eutectic temperatures shows in the case of Ni and Co a correlation between both. In comparison with this, the siliconizing of Mo is only possible at low rates in the given temperature range.

At lower temperatures silicide formation by solid-state diffusion was found and at higher temperatures a solidification structure with all materials studied. In the case of Co it was found that, analogously to Ni, a Co_3Si–Co structure is formed. The structures of the layers formed on

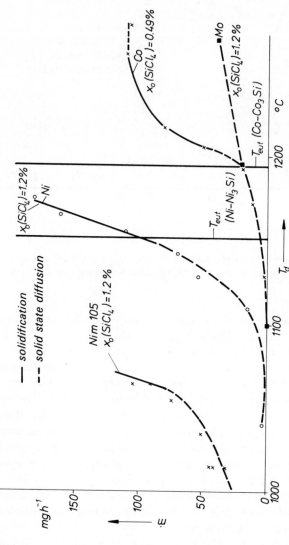

FIG. 10. Dependence of the siliconizing process on the substrate material—the experiments were performed in the cold wall reactor: $p_{tot} = 200$ torr, $i_{tot} = 33$ l/h (STP)

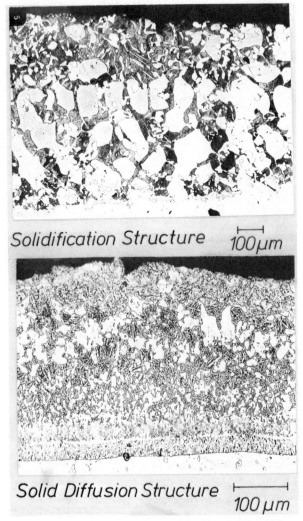

FIG. 11. Section of siliconized layers on Nim 105

Nim 105 are shown in Fig. 11. The silicides found in the layers should be compared with the results of Felix and Erdös (4). Electron microprobe analyses of the silicide layers produced by solid diffusion show that there is Cr enrichment (4) near the substrate boundary and Cr depletion in the outer zones. This effect could be caused by Cr evaporation (see Fig. 3) and/or by a selective siliconizing process, because the NiSi compounds are more stable than the CrSi compounds (10). Silicide layers produced by solidification from the melt (Fig. 11a) showed no concentration profile through the layer.

At low Si contents in the gas and at high temperatures, Si/solid solution layers are produced on the superalloys, as in the case of Ni. The deposition conditions for the formation of the various layers on IN-738

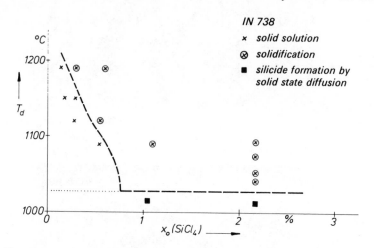

FIG. 12.   Boundaries between the siliconizing ranges for IN-738. The experiments were performed by glow discharge siliconizing

(8.5Co/16Cr/3.5Al/3.5Ti/1.6Ta/2.5W/1.8 Mo, Ni balance, wt%) are shown in Fig. 12. A typical diffusion profile is demonstrated in Fig. 13.

Assuming diffusion of Si only, with a concentration independent diffusion coefficient D_{Si}, the profile of the silicon concentration C_{Si} can be described by the following equation (Ch. II, sec. 2.9 in ref. 14):

$$C_{Si} = \frac{2j_0}{\sqrt{D_{Si}}} \sqrt{(t)} ierfc\left(\frac{x}{2\sqrt{(D_{Si}t)}}\right) \qquad (2)$$

where t is the siliconizing time, x the distance from the surface, and j_0 the deposition rate of silicon atoms on to the alloy surface.

The profiles calculated from eqn. (2) compare well with the experimental profiles shown in Fig. 13. It is useful to describe the silicon penetration by the distance $d_{0.1}$ from the surface at which the silicon concentration has decreased to the value $C_{Si} = 0.1 C_{Si_0}$ (C_{Si_0} is the concentration of Si at the

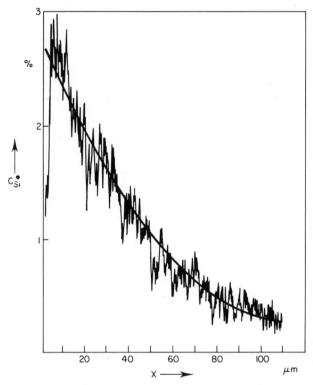

FIG. 13. Silicon/concentration profile on IN-738; the profile was measured by microprobe analysis: $x = 0$, surface of the sample; C_{Si}^*, mole fraction of Si; —— calculated by eqn. (2)

surface). With the help of eqn. (2), $d_{0.1}$ can be calculated by

$$d_{0.1} = 2\sqrt{(D_{Si}t)} \qquad (3)$$

In Fig. 14 the penetration depths, measured on the materials investigated, are plotted against the reciprocal absolute temperature. Additionally the diffusion coefficient calculated from eqn. (3) is displayed. According to Fig. 14 the diffusion coefficient can be described by

$$D_{Si} = D_0 \exp(-E/RT) \qquad (4)$$

where $D_0 = 0.5\,\mathrm{cm^2/s}$ and $E = 62\,\mathrm{kcal/mol}$, and is approximately independent of the material investigated (Ni, IN-738, Nim 105). The measured activation energy is in good agreement with the value

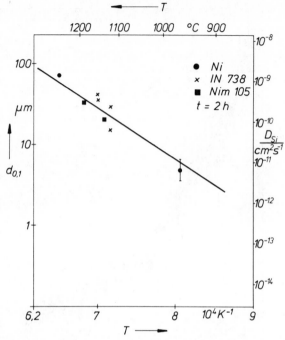

FIG. 14. Measured penetration depths $d_{0.1}$. Straight line: calculated by eqns. (3) and (4). Additionally, the diffusion coefficient, eqn. (3), is plotted on the vertical axis. The penetration depths $d_{0.1}$ were measured after a siliconizing time $t = 2\,\mathrm{h}$. The point ⬢ is calculated from a measured Si profile on a Ni sample heated for 500 h

($E = 61.7\,\mathrm{kcal/mol}$) measured in ref. 15 for the diffusion of Si in Ni, whereas the pre-exponential factor found in ref. 15 is larger ($D_0 = 1.5\,\mathrm{cm^2/s}$).

According to eqn. (2) the silicon concentration of the surface is given by

$$C_{Si} = 2j_0 \sqrt{\left/\left(\frac{t}{\pi D_{Si}}\right)\right.} \tag{5}$$

For simple cases, where the phase diagrams for the systems MeSi (Me = metal) are well known, the boundaries of the solution range can be calculated by eqn. (5). In this way the boundary between the melting range and the solid-solution range is calculated for the siliconizing of Ni. The calculated boundary is in good agreement with the experimental data as shown in Fig. 9. The curved form of the boundary is caused by the combination of the temperature dependence of the solidus line with that of the diffusion coefficient.

DISCUSSION

Comparison of the thermodynamic calculations and the experimental results shows that in the hot wall reactor as well as in the cold wall reactor the silicon transport is determined by kinetic processes which have a very strong temperature dependence, in contrast to the weak effect of temperature shown by the thermodynamic calculations (Fig. 3). The dependence of the process on the composition of the substrate material (Fig. 10) and on the molecules in the gas phase ($SiCl_4$, $SiHCl_3$) leads us to presume that surface processes are the rate-controlling steps. The different deposition rate in the hot wall and the cold wall reactors can be explained by the different rates of approach of the gas atmospheres towards thermodynamic equilibrium by the homogeneous reactions in the two cases. These reactions are less likely in the cold wall reactor because the gas is heated to the deposition temperature T_d only in the boundary layer (thickness ≈ 1 mm (16)) near the sample. In the hot wall reactor, however, there is no temperature boundary layer because the whole reactor is heated to the deposition temperature T_d.

The high deposition rate and the very weak temperature dependence in the gas discharge suggest that the deposition is not controlled by chemical reactions with a high activation energy but by other processes, e.g. diffusion processes in the gas phase. The weak temperature dependence can then be explained by the weak temperature dependence of the diffusion coefficients (17). The increased deposition rate in the glow discharge is in accord with the results of Koenings *et al.* (18), which showed that the oxidation of $SiCl_4$ to SiO_2 can be enhanced by a microwave discharge.

The result of this investigation into the different siliconizing processes enables us to minimize the temperature dependence of the silicon transfer to the material to be siliconized. The temperature dependence of the solid-state diffusion processes, however, which form the silicon-enriched layers, cannot be affected because it is determined by the fixed activation energy of the diffusion coefficient.

ACKNOWLEDGEMENTS

The authors are indebted to Mr W. Sick and Mr D. Gilbers for assistance with the experimental work and to Mr A. Nicoll who assisted with the English manuscript.

REFERENCES

1. G. V. Samsonov et al., Eds. (1966). *Coatings of High Temperature Materials*, Plenum Press, New York.
2. E. Fitzer (1954). *Arch. Eisenhütt.Wes.*, **25**(11/12), 601.
3. E. Fitzer (1952). *Bergb. Hütte. Monatsh.*, **97**(5), 81.
4. P. Felix and E. Erdoes (1972). *Werkstoffe Korros.*, **23**(8), 627.
5. P. R. Sahm (1976). *Metallk.* **30**, 326.
6. P. C. Felix and H. Beutler (1972). *Third Chemical Vapor Deposition Proceedings* (Ed. F. A. Glaski), ANS, Hinsdale, Ill., p. 600.
7. L. P. Hunt and E. Sirtl (1972). *J. Electrochem. Soc.*, **119**, 1741.
8. I. Barin and O. Knacke, Eds. (1973). *Thermochemical Properties of Inorganic Substances*, Springer, Heidelberg.
9. D. R. Cruise (1964). *J. Phys. Chem.*, **68**, 3794.
10. T. G. Chart (1972). 'A Critical Assessment of Thermochemical Data for Transition Metal–Silicon Systems', NPL Report Chem. 18 (August).
11. A. V. Engell (1965). *Ionized Gases*, Clarendon Press, Oxford.
12. M. Hansen (1958). *Constitution of Binary Alloys*, McGraw-Hill, New York.
13. F. C. Eversteyn (1974). *Phillips Res. Reps.*, **29**, 45.
14. H. S. Carslaw and I. C. Jaeger (1959). *Conduction of Heat in Solids*, Clarendon Press, Oxford.
15. R. A. Swalin, A. Martin and R. Olson (1957). *J. Metals*, **9**, 936; cited in Landolt Börnstein (1968). *Zahlenwerte und Funktionen*, II. Band, 5. Teil, Bandteil b, Springer, Heidelberg.
16. H. Schlichting (1965). *Grenzschichttheorie*, Braun, Karlsruhe.
17. R. C. Reid and Th.-K. Sherwood (1966). *The Properties of Gases and Liquids*, McGraw-Hill, New York.
18. J. Koenings, D. Küppers, H. Lydtin and H. Wilson (1975). *Fifth Chemical Vapor Deposition Proceedings* (Eds. J. M. Blocher, H. E. Hintermann and L. H. Hall), ECS, Princeton, N.J., p. 270.

DISCUSSION

R. Pichoir (ONERA, 92320 Chatillon, France): When you heat your specimen by ion bombardment, the coating is being sputtered during its formation. What is the importance of sputtering in your case? Do you take this factor into account in determining the siliconizing kinetics of the various alloys?

Reply: In the cathode drop region near the sample, positive ions are accelerated to the surface of the sample. They have an energy U of approximately $300 \, eV < U < 600 \, eV$ and a current density, j, $10 \, mA/cm^2 < j < 50 \, mA/cm^2$. The ions—mainly hydrogen ions—

produce the following effects by impingement on to the surface of the cathode: (1) heating of the sample to temperature, and (2) sputtering of atoms and ions from the surface into the gas phase.

In order to measure the effect of sputtering, weight change measurements were made without silicon in the gas phase. In these experiments the heat treatment of the sample in the glow discharge was similar to that used during the siliconizing process (Fig. 6). In these experiments we found weight losses Δm caused by sputtering processes (and thermal evaporation processes) which were negligible ($< 10\%$), in comparison with the weight gain during the siliconizing process. Therefore we neglected this effect in our experiments.

R. Pichoir: Do you observe a dissociation of the gaseous phase containing silicon within the plasma, leading to the formation of finely divided silicon powder?

Reply: The composition of the gas phase in the glow discharge has not yet been analysed, but dissociation processes are very probable. In addition, highly excited atoms and molecules are formed. On the heated surface powder formation was not observed in any of the siliconizing processes. But on the cold walls of the reactor we found a powder deposit consisting of Si and small amounts of NiCr from the siliconized sample. This powder is possibly formed by homogeneous nucleation processes.

H. Pflug (Robert Bosch GmbH, Schweberdingen, Germany): Have you also carried out corrosion tests on pure nickel protected by coatings rich in silicon?

Reply: We have performed corrosion tests on silicide layers in air (1000 °C, 100–1000 h). The silicide layers were only produced in the melting range (Fig. 6) because the silicide layers produced by solid-state diffusion were very brittle and the adhesion to the basic material was bad. Silicon-enriched layers in the solid solution range have yet to be tested. The silicide layers tested were stable against corrosion in air. Silicon depletion was found to be due to Si diffusion (see Fig. 14 and paper of U. W. Hildebrandt *et al.*, Chapter 14).

21

Some Aspects of Silicon Coatings Under Vanadic Attack

P. Elliott and T. J. Taylor

UMIST, Manchester, UK

ABSTRACT

The influence of molten vanadium-rich deposits upon silicon-coated Ni/20Cr alloy has been examined at 900 °C under oxygen using 80–20 % mixtures of V_2O_5 and Na_2SO_4 as corrodant. Coatings applied by pack, vapour deposition, plasma spraying and ion plating were all effective, with catastrophic rates for uncoated material, reduced by as much as 80 % for coated material. Ion-plated coatings were found to be the most reliable with good adhesion, uniform composition and even thickness. Tests in the laboratory have shown no breakaway effects for times in excess of 600 h at 900 °C.

Increased protection appears to be associated with the development and retention of a Cr-rich barrier layer beneath the Si-rich surfaces and there is evidence to show that the experimental slags show little reaction with Si.

INTRODUCTION

Interest in corrosive attack associated with molten vanadium-rich deposits arising from the combustion of heavier fuel oils remains, despite the claims that such corrosion can be controlled or minimized by a combination of additives, coatings and material development (1). The problem of vanadic corrosion is not new and is associated with the combined presence of vanadium (arising as a porphyrin), sodium (mainly from sodium chloride), sulphur (from the fuel) and oxygen. During combustion, vanadium oxides and sodium sulphate can condense as low melting vanadyl vanadates which permit rapid corrosion, at temperatures of ~ 600 °C, by removal of normally protective surface oxides. The problem is particularly troublesome in high

353

temperature applications, such as furnace parts or turbine blades, where cheaper but less pure residual or lower distillate fuel oils are utilized.

Work on reducing the severity of vanadic corrosion has been hindered by a lack of information about the chemistry of the corroding system. As will become apparent in this chapter, certain physicochemical properties of the slag mixtures need to be reconciled if improved lifetimes of components are to be achieved.

This chapter describes results obtained from silicon-coated Ni/20Cr alloy exposed to simulated vanadic slags based on V_2O_5–Na_2SO_4 (80–20%). Various coating techniques were tried, although during the course of the work particular promise was found with ion-plated silicon which did not suffer from the defects of other techniques.

Laboratory testing is often frowned upon because of the difficulty of predicting behaviour in practice. The authors are aware of the dangers inherent is such cases and care was taken to ensure reasonable reproducibility in testing. Relatively simple tests have been employed in order to investigate the effects of silicon as a coating barrier to corrosion. The tests reported herein relate to metal-excess situations where small quantities of simulated slag are contained in holes or cup-shaped samples. Such techniques have been useful in previous studies (2, 3) and have the advantage of exposing slag-free surfaces simultaneously with slag-contacted surfaces. All tests were conducted in flowing oxygen at about 1 atm pressure.

EXPERIMENTAL

Two types of test specimen were used: 0.5 mm thick dish samples capable of holding ~ 50 mg of slag, and heavier block samples ($10 \times 10 \times 5$ mm) with 2.5 mm dia. flat-bottomed holes drilled into them. Samples were polished to a 600 silicon carbide grit finish, degreased ultrasonically in methanol and acetone and stored in vacuum desiccators until required. The Ni/20Cr alloy, supplied by INCO(Europe) Ltd had the following composition: Ni 79.23%, Cr 20.6%, Mg 0.023%, Si <0.1%, Co, Ti, Mn, Mo, Al, Cu <0.05%, C <0.002%, all by weight.

Coatings
Coatings were successfully produced using several types of technique, despite certain limitations associated with each method. However, it should

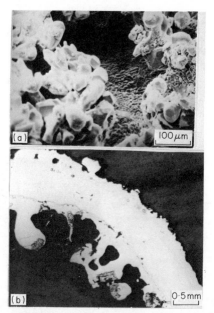

FIG. 1. SEM and cross-section through Si pack-coated Ni/20Cr (1 h at 1000 °C)

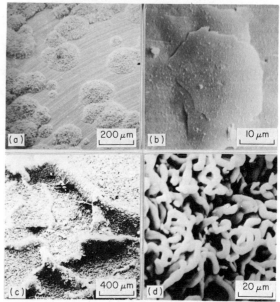

FIG. 2. SEM of vapour-deposited Si coatings on Ni/20Cr: (a, b) SiCl₄ at − 30 °C
for 1 h; (c, d) SiCl₄ at 30 °C for 1 h

356   *P. Elliott and T. J. Taylor*

be mentioned that the primary objective in this study was to assess whether silicon was a useful coating material under vanadic attack.

Preliminary studies with *pack coatings* resulted in very irregular coating thicknesses and generally poor adhesion with non-continuous coverage. These features are demonstrated in Fig. 1 which shows a typical sample coated for 1 h at 1000 °C under dried argon. Packs were 1 % NH_4Cl, 25 % Si, bal SiO_2 to 60 mesh particle size (250 μm).

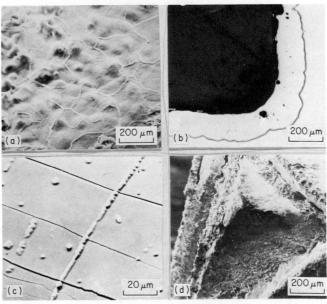

FIG. 3.   Vapour-deposited Si coatings on Ni/20Cr; $SiCl_4$ at 0 °C for 1 h: (a, b) SEM and cross-section; (c) fine cracks in coating; (d) detached corner

Coatings produced by *vapour deposition* from silicon tetrachloride were more promising, although several trials were necessary to achieve optimum temperatures and conditions for coating. Too low a halide temperature produced only nuclei of silicon (Fig. 2a) or too thin coatings (~ 2–3 μm) which were often cracked (Fig. 2b). Too high temperatures resulted in ridged or worm-like configurations indicative of vaporization/condensation effects (Fig. 2c, d). Figure 3 shows that reasonably uniform coatings could be obtained if conditions were carefully adjusted. Fine cracks were found at higher magnifications (Fig. 3c) and adhesion was not exceptionally strong since corners could easily be detached from cold samples (Fig. 3d).

Plasma-sprayed coatings on shot-blasted surfaces gave good coverage, even within the drilled-out holes but such coatings were porous and adhesion was weak in places. Figure 4(a) shows the general roughness of a plasma coating applied for ~ 1 s on to a specimen rotating at ~ 80 rev/min. The cross-section in Fig. 4(b) shows the extent of inhomogeneities and porosity associated with the technique.

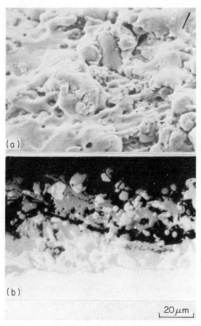

20 μm

Fig. 4. Plasma-coated Si in SEM (a) and in cross-section (b)

Ion-plating incorporates a good throwing power so that all surfaces can be coated simultaneously and with considerable uniformity and adhesion (Fig. 5a). Coating thicknesses could be varied between 7 and 20 μm by controlling plating conditions, particularly time and potential, and the surface concentration of Si by electron probe microanalysis (uncorrected) was $\sim 75\%$. Some 'splashing' effects were observed on occasion (Fig. 5b) but such defects should be avoidable with refinements to the coating procedures. It is interesting to note that a silicon-rich layer is still present beneath such features, as can be noted in a particularly large 'splash' shown in Fig. 5(c, d).

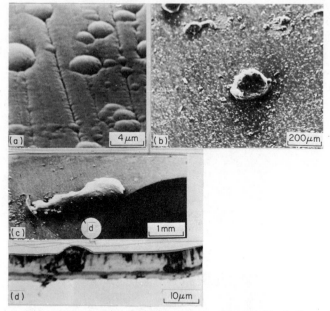

FIG. 5.   Ion-plated Si: (a) typical SEM appearance; (b) 'splashing' effects; (c) large
'splash' of Si; (d) cross-section beneath (c)

Corrosion Testing

Tests were conducted for various times at 900 °C or above where the vanadic
slags would be molten. Corrosion rates were assessed continuously in a C.I.
Electronics automatic thermobalance or by weight loss estimates after
testing in horizontal controlled atmosphere rigs. For comparison purposes
uncoated specimens were exposed to identical corrosion or oxidation
environments and coated specimens were oxidation tested to provide a
baseline for assessing their protective behaviour. This chapter is only
concerned with tests in 1 atm oxygen.

RESULTS

Kinetics

Figure 6 contrasts oxidation behaviour with vanadic corrosion for
Ni/20Cr, coated or uncoated, when exposed to 1 atm oxygen at 900 °C for
100 h. Oxidation rates were low, as would be anticipated for Cr-rich surface

films on Ni/20Cr or SiO$_2$-type products on the coated alloy, with overall weight gains ~ 1–2 mg/cm^2 in the 100 h period.

The presence of the vanadic slag resulted in an almost linear kinetic response as is anticipated for such environments; weight gains of ~ 40–50 mg/cm^2 were common. The slight decrease in corrosion rate shown in Fig. 6 could be associated with saturation of the melt by the corrosion products.

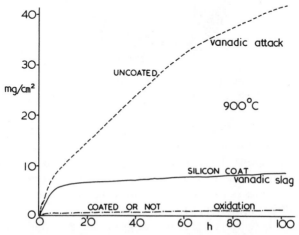

FIG. 6. Kinetics of oxidation and vanadic corrosion for Ni/20Cr with and without coatings at 900 °C in oxygen atmosphere

The presence of a silicon coating resulted in a marked reduction in kinetics with a more protective tendency (Fig. 6). Weight gains after 100 h were ~ 8 mg/cm^2 for ion-plated samples, the rate of attack being lower than for plasma- or pack-coated systems.

Extent of Corrosion
Conventional optical and electron optical microscopy were useful in determining the extent and form of attack. As is already known, vanadic slags give rise to very rapid attack with resultant complex morphologies, as can be judged from the degree of attack shown by the Ni/20Cr samples in the present experimental tests. Oxidation at 900 °C is a relatively slow process, there being ample protection from Cr-rich surface oxides (Fig. 7a). By contrast, the metal has been wholly consumed in certain sections when exposed to the vanadic slag for the same period of 150 h (Fig. 7b).

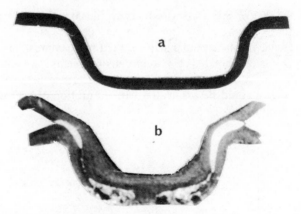

FIG. 7. Sections through Ni/20Cr after 150 h at 900 °C under oxygen: (a) oxidation test (dark field illumination); (b) vanadic slag test

Cross-sections (e.g. Fig. 8) reveal various zones, comprising an inner amorphous layer (Fig. 8a), an intermediate fine duplex layer (Fig. 8b) and an outer course duplex layer (Fig. 8c). Attack generally occurs on a broad front in a uniform manner (Fig. 8d), although localized grain boundary attack by sulphides is found on occasion (Fig. 8e) despite the oxidizing character of the surrounding test atmosphere.

An increase in temperature is also known to increase vanadic corrosion (4) and this is clearly apparent in the situation at 1000 °C after 20 h, where extensive attack has given rise to voluminous products (Fig. 9).

Associated with such conditions was the presence of many laminations parallel to the corroding surface which might be indicative of some form of repeated fluxing mechanism because Cr-rich bands were revealed by electron probe analysis.

The presence of Si has a dramatic effect upon the behaviour of Ni/20Cr alloy in vanadic slags based on V_2O_5 and Na_2SO_4. The bulky scales present without the coating are absent and the degree of metal loss is much reduced; typically, an 80 % saving on metal loss compared with the uncoated alloy. An important feature in the protection mechanism is the diffusion of Si into the alloy which resulted in the formation of a Si-rich NiCr surface layer which shows relatively little reaction with the vanadic environment.

Uncoated samples gave products dominated by Ni-rich vanadates (Fig. 10a), whereas Si-coated samples gave rise to products containing Ni, Cr, Si and V. Vanadates rich in Cr and Si along the outer slag layers were often observed (Fig. 10b). This complex band of vanadates on the outer surfaces

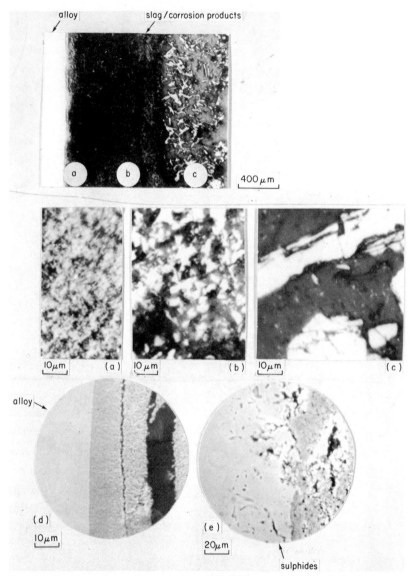

FIG. 8.   Features of vanadic corrosion of Ni/20Cr revealed by optical microscopy after testing at 900 °C: (a) inner amorphous layer; (b) intermediate duplex layer; (c) outer coarse duplex layer; (d) detail of metal interface—uniform attack; (e) detail of metal interface—local intergranular sulphides

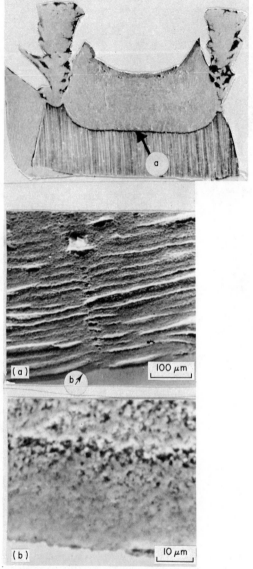

FIG. 9.   Vanadic corrosion of Ni/20Cr at 1000 °C for 20 h: (a) laminations in products; (b) metal interface

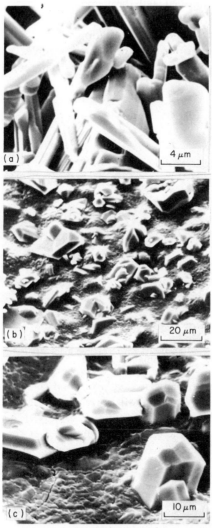

FIG. 10.   SEM photographs of Si-coated Ni/20Cr (ion plated) after vanadic corrosion at 900 °C for 100 h: (a) Ni-rich vanadates for uncoated Ni/20Cr; (b) modified vanadates for Si-coated Ni/20Cr; (c) Si-rich matrix with outcrops of Cr-rich products

was associated with an inner Cr-rich band of material which provides additional resistance to vanadic attack. With time, the complex surface vanadates are diminished and the Cr_2O_3-rich inner layer becomes less defined. Ultimately, vanadic corrosion can become more significant but not so extensive as for the uncoated situation.

The general behaviour for the coated samples was of protection, provided that the integrity of the coating was maintained—the integrity of

FIG. 11.   Ballooning of residual poured slag after reaction with immersed Si sample for 0.5 h at 750 °C

the ion-plated system being particularly good. Detailed examination of corroded samples showed a smooth Si-rich layer occasionally disturbed by protrusions of oxide rich in Cr containing some V (Fig. 10c). This type of situation suggests that the vanadic slag did not 'wet' the Si-rich coating or that the Si itself did not have significant reaction in the melt.

It is interesting to note also that certain physical changes were observed when small samples of pure Si, Cr, Ni or Ni/20Cr alloy were dipped into molten slag for 0.5 h at 750 °C under oxygen. Upon removal all samples showed weight changes and reaction, with the exception of Si which was little affected (Table 1). The residual melts were each poured on to a flat alumina surface after test and all settled down to a low flat profile with the exception of that which had contained the Si which ballooned up into a hollow sphere (Fig. 11). This behaviour suggests that enhanced gas (oxygen) evolution had occurred on cooling which infers that little oxygen

TABLE 1

Immersion tests at 750 °C for 0.5 h in vanadic slag under oxygen

Metal	Weight change (%)	Profile of poured slag
Ni	55.5 (loss)	flat
Cr	1.5 (gain)	flat
Ni/20Cr	5.4 (gain)	flat
Si	0.4 (gain)	balloon (see Fig. 11)

had been consumed during the melt immersion with Si compared with the other metals. It is known that V-rich slags containing alkali metals absorb oxygen on melting and liberate the same amount of oxygen on freezing (5), so it would appear that some other effects are present when metallic additions are made.

DISCUSSION

The morphological features described herein give some indication of how complicated the mechanism of corrosion by molten vanadic slags can be. The process involved is really an accelerated oxidation phenomenon which will be affected by many variables including temperature, partial pressure of oxygen, melt composition and depth.

In the case of vanadic corrosion, up to ten separate or interrelated steps may be rate-determining at some point in the process (3, 6). Molten V_2O_5 has been described as a transition semiconductor (5) which implies that any added species, especially cations, would invariably increase electrical conductance and hence the corrosion rate. What could be important are the secondary effects such as increased melting point products, reduced diffusivity properties or changes in fluid properties, which can occur with various levels of additive (7).

The presence of Si has been beneficial for overcoming certain problems associated with deposits and polluted gases (1, 8) and some earlier observations (9) suggest that Si, as SiO_2, appeared to have better resistance to attack by molten V_2O_5 whereas stable Cr_2O_3-type oxides were destroyed, which is somewhat contrary to the generally held view that Cr is perhaps the most useful element for resisting hot corrosion.

The present study has clearly demonstrated that Si-type coatings are of particular benefit in laboratory tests with V_2O_5–Na_2SO_4 melts on Ni/20Cr under oxygen (recent tests have shown this benefit to continue to times in excess of 600 h at 900 °C).

The physical effects observed in the present study suggest that interactions with Si rather than SiO_2 are more significant, in particular the formation of Si-rich, Ni/Cr layers on the alloy surface. The presence of the alkali metal sulphate should also not be overlooked, especially as the uncoated samples showed signs of localized intergranular sulphidation despite the high partial pressure of oxygen above the melt.

What emerges from the study so far is the fact that Si promotes the formation of products dominated by complex NiCrSi vanadates which are apparently beneficial in supporting Cr-rich sublayers. It is known that Cr (and Ni to a less extent) can form barrier layers within vanadic melts (5), (i.e. the metal oxidation products dissolve in the melt slower than they form), which could help explain the effectiveness of the present results.

If vanadic corrosion is accepted as a diffusion-controlled process (5), then it is possible to envisage two operative sequential processes: (1) the inward diffusion of oxygen via the melt and/or barrier layer, and (2) the transport of corrosion products away from the metal, which will vary from metal to metal. The presence of Si permits the formation of a different form of oxide or vanadate on the surface which hinders the rapid removal of normal scales by fluxing or reaction with oxygen arriving through the melt. Furthermore, there is evidence that Si and V_2O_5–Na_2SO_4 slags do not react in the short term (Fig. 11 and Table 1) nor the longer term tests (Fig. 10c), which could be related to wetting and/or adhesion effects rather than diffusivity or transport phenomena. Associated with this is the suggestion that Si and V do not apparently form compounds with each other (10).

CONCLUSIONS

(1) Si coatings can be an effective means of reducing vanadic corrosion of Ni/20Cr by V_2O_5–Na_2SO_4 melts under oxygen.

(2) Barrier-type layers involving both Si- and Cr-containing species are apparently associated with the improved behaviour.

(3) Ion plating of Si affords a more reliable form of coating, although the various coating techniques examined have all been effective to some extent.

ACKNOWLEDGEMENTS

The authors are grateful to SRC and INCO(Europe) Ltd for research funding to T. J. T. The co-operation of the following is much appreciated:

Dr D. Teer (University of Salford) for ion-plating work, Dr F. R. Sale (University of Manchester) for assistance with chemical vapour deposition, and INCO(Europe) Ltd for provision of materials and plasma-spray coating work.

REFERENCES

1. K. J. Williams and P. J. Parry (1972). 'High Temperature Corrosion of Aerospace Alloys'. Conf. Proc. No. 120, AGARD, Tech. Univ. of Denmark, Lyngby.
2. P. Elliott and A. F. Hampton (1973). *Deposition and Corrosion in Gas Turbines* (Eds. A. B. Hart and A. J. B. Cutler), Applied Science, London, Ch. 15, p. 244.
3. R. B. Dooley, R. C. Kerby and J. R. Wilson (1971). 'High Temp. Corrosion of Materials in Fuel Combustion Products', Prog. Rept 2, DRBG grant 7535-14; see also *Can. J. Chem.*, **50** (1972), 2865–71.
4. P. Elliott, T. K. Ross and G. C. Soltz (1973). *J. Inst. Fuel*, **46**, 77.
5. D. A. Pantony and K. I. Vasu (1968). *J. Inorg. Nucl. Chem.*, **30**, 755.
6. D. A. Pantony and K. I. Vasu (1968). *J. Inorg. Nucl. Chem.*, **30**, 423.
7. D. A. Pantony and K. I. Vasu (1968). *J. Inorg. Nucl. Chem.*, **30**, 433.
8. J. Graham (1974). Conf. on 'Techniques to Minimize High Temperature Corrosion by Protective Coatings, Additives and Fuel Treatments' (European Fed. of Corrosion, Copenhagen, May).
9. E. Fitzer and J. Schwab (1956). *Corrosion*, **12**, 49.
10. W. R. May, M. J. Zetlmeisl and R. R. Annand (1973). ASME Publ. 73-WA/CD-1.

22

Hot-corrosion Behavior of Chromium Diffusion Coatings

R. Bauer, H. W. Grünling and K. Schneider

Brown Boveri & Cie, Mannheim, West Germany

ABSTRACT

This chapter presents the results of full-scale engine testing of first-stage turbine blades with a chromium diffusion coating.

A mechanism involving outward chromium diffusion from a chromium-rich reservoir layer through an overlying working layer with constant chromium content is proposed to account for the predominantly diffusion-controlled consumption rate of the coating. As a consequence, accelerated corrosion of the coating was observed at higher temperatures due to the increase in the rate of oxidation leading to accelerated chromium transport.

Correlations have been made with laboratory crucible tests under more severe conditions which have indicated initial rapid corrosion degradation processes due to progressive chromium depletion and the subsequent formation of fewer protective mixed oxides.

INTRODUCTION

The chromium-rich coatings used to increase the service life of stationary gas-turbine blades are today one of the most developed protective systems against high temperature corrosion attack (1–4). Chromium diffusion coatings produced by the various pack cementation techniques are in particular finding widespread applications (5, 6).

Here, the workpiece to be coated is packed in a powdered mixture of the coating element (Cr), a small amount of an easily decomposable activator (e.g. NH_4Cl) to produce the gas phase, and an inert ballast material (normally Al_2O_3) to prevent sintering. Reaction is carried out under either an inert gas or hydrogen atmosphere at elevated temperatures (e.g.

369

800–1100 °C) for some hours, depending on the nature and thickness of the coating required.

The protective mechanism of such coatings is analogous to that of the chromium-rich superalloys, and depends on their ability to develop a dense, coherent oxide coating (Cr_2O_3) as a diffusion barrier against further oxidation or sulfidation. As a consequence of the interaction of hot gases and fuel-ash deposit, such coatings undergo a continuous consumption which, in addition, is accelerated by thermal and mechanical stressing. For example, creep stress above a critical limit may cause cracking of the protective oxide scale to occur at a rate faster than the rate of regrowth of the oxide. This is also true for the spalling and subsequent oxide reformation due to stresses resulting from thermal cycling.

The progressive chromium impoverishment at first causes protective oxide phases to form and ultimately the chromium content becomes insufficient for their formation. Together with this are the chemical and physical interactions between the base material and its coating which determine the properties of the protective scale produced, and may also cause significant changes in these during service. Thus, the chromium transport mechanism plays an important role in the life expectancy and ultimate failure of the protective system.

Despite the widespread use of such coatings there is little information available regarding their behavior either in service or during corrosion testing, and above all concerning the causes of the chemical interactions involved and their effect on the mechanism of protection and the coating life expectancy (7, 8). As a consequence of this we have attempted to describe the thermochemical processes within chromium diffusion coatings on the forging alloy Nimonic 80A, and the casting alloy Inconel 738 LC.

The program of research included investigations of: (a) the coating as produced; (b) the coating after gas turbine service at 720–750 °C and 850 °C gas inlet temperatures, respectively; and (c) the coating after laboratory crucible testing at 850 °C in a synthetic slag.

THE COATING AS PRODUCED

Figure 1 shows the distribution of chromium, iron (from the coating process) and nickel (from the base metal) across a typical chromium diffusion coating. Both the diffusion zone and growth zone above the original metal surface are clearly visible. The outer two-phase growth zone contained chromium, iron and nickel together with small concentrations of the outward diffusing base metal elements Co, Mo and W, but noticeably

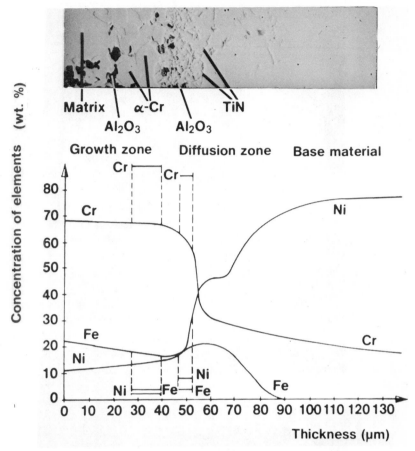

FIG. 1. Chromium diffusion coating on Nimonic 80A after heat treatment

no Al. The growth zone consists of a matrix of 60–70 % chromium and nickel, with the latter decreasing slightly in concentration towards the outer surface. At higher magnifications the matrix proves to contain a second phase in the form of a needle-shaped precipitate. Embedded in this matrix is an α-chromium precipitate with the typical composition 90 % Cr, 5 % Fe and 5 % Ni. Globular oxide particles are also present, these proving to be the ballast material from the diffusion coating process (in this case corundum).

The diffusion zone is characterized by a steady decrease in concentration of the coating element chromium together with an iron-rich γ-phase leading

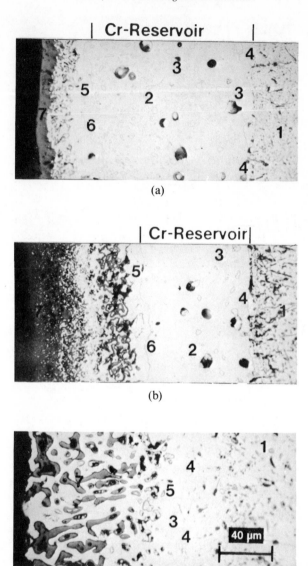

FIG. 2. Chromium diffusion coating on Nimonic 80A after 35 000 h service at 710–730 °C; (a) little attack, (b) severe attack, (c) residual coating. $1 = \gamma\text{-Ni}_{0,3}\text{Fe}_{0,4}\text{Cr}_{0,3}$. $2 = \sigma\text{-Fe}_{0,38}\text{Cr}_{0,50}\text{Ni}_{0,12}$. $3 = \gamma\text{-Ni}_{0,3}\text{Fe}_{0,4}\text{Cr}_{0,3}$. $4 = \alpha\text{-Cr}_{0,8}\text{Fe}_{0,15}\text{Ni}_{0,05}$. $5 = \gamma\text{-Ni}_{0,2}\text{Fe}_{0,6}\text{Cr}_{0,2}$. $6 = \sigma\text{-Fe}_{0,47}\text{Cr}_{0,46}\text{Ni}_{0,07}$. $7 = \text{Cr}_2\text{O}_3$

into the original composition of the base material. Within the region adjacent to the base metal surface, apart from chromium carbide, Al_2O_3 and a titanium carbonitride precipitate, may be observed. These arise from the selective oxidation and nitriding, respectively, of the γ'-phase during coating.

CONDITION OF THE COATING AFTER SERVICE

The following observations were made on a coated Nimonic 80A first-stage stator vane (gas-inlet temperature 710–730 °C) from a blast furnace gas turbine based in the Ruhrgebiet. The turbine afforded only a relatively mild corrosive environment in comparison with other machines of this class. After 35 000 h service no visible signs of corrosion attack could be determined. For comparison, a second uncoated vane of the same material, mounted in the adjacent position, showed clear evidence of corrosion.

Figure 2(a–c) shows varying thicknesses of the coating remaining after 35 000 h service. In all cases a structure of clearly defined zones is observed. Beneath a compact scale of chromium oxide is a metallic zone partly interspersed with the chromium oxide. Between this and the base material itself one finds a clearly defined middle zone exhibiting varying stages of consumption, the outer portion consisting of two and the inner of three phases. The metallic zone grows inwards together with the outer oxide (Fig. 2a,b) until finally only isolated islands of this middle zone remain (Fig. 2c).

Figure 3 shows the distribution of chromium, nickel and iron across the region shown in Fig. 2(a). The specific oxide and metallic precipitates have not been taken into account.

From electron microprobe analyses and X-ray diffraction the following results were obtained:

(1) The dense outer oxide scale, together with the numerous oxide protrusions penetrating into the outlying metallic zone, consists of nearly pure chromium oxide (Cr_2O_3).

(2) The outlying metallic zone is composed of an iron-rich γ-ternary solid solution with a constant composition of 56 % Fe, 23 % Ni and 21 % Cr.

(3) The matrix of the middle zone consists of the σ-phase (structure type $D8_b$; space group $P4_2/mmm$; tetragonal, with $a_0 = 8.78$ Å; $c_0 = 4.556$ Å) having a composition of 47 % Fe, 46 % Cr and 7 % Ni in the $\sim 10\ \mu m$ layer forming the 'boundary zone' to the outlying γ-phase; and of 50 % Cr, 38 % Fe and 12 % Ni in the remainder.

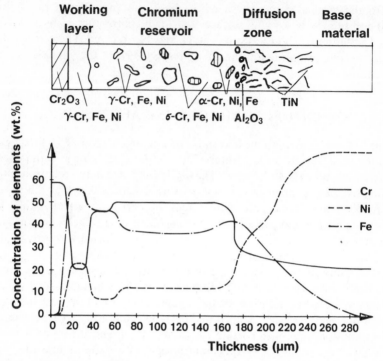

FIG. 3. Chromium diffusion coating on Nimonic 80A after operating test; distribution profiles of elements and phases

(4) The light-coloured precipitate in the σ-phase consists of a γ-solid solution with a composition of 43% Fe, 33% Ni and 24% Cr.

(5) The dark-coloured precipitate in the inner half of the σ-zone appears to be an α-chromium phase with $\sim$82% Cr, 15% Fe and 3% Ni.

(6) The gradient of the diffusion zone in the base metal is steep in accordance with the observations of compound formation. The iron enrichment in the inner half of the σ-zone corresponds to a γ-solid solution of composition 40% Fe, 32% Ni and 28% Cr. The diffusion zone has a thickness of about 120 μm.

(7) The continuous layer interdispersed with chromium oxide (Fig. 2c) has a composition similar to that of the outer γ-zone.

(8) Sulfur in the form of randomly dispersed chromium sulfides was found at the tips of the penetrating oxide pegs.

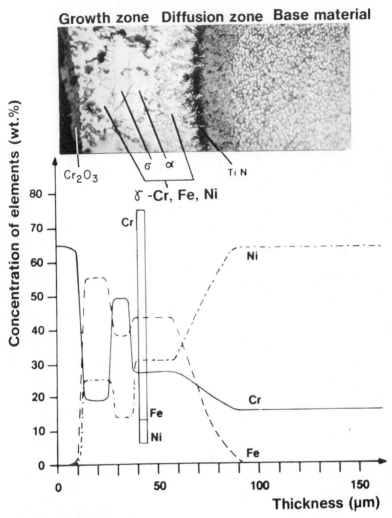

FIG. 4.   Chromium diffusion coating on IN-738 LC after operating test at ~750°C

(9)   Oxide inclusions originating from the chromizing powder were not found (cf. Fig. 1).

Comparable distributions of these elements have also been observed across chromium diffusion coatings on IN-738 LC blades after service in other gas turbines at temperatures of ~750°C (Fig. 4).

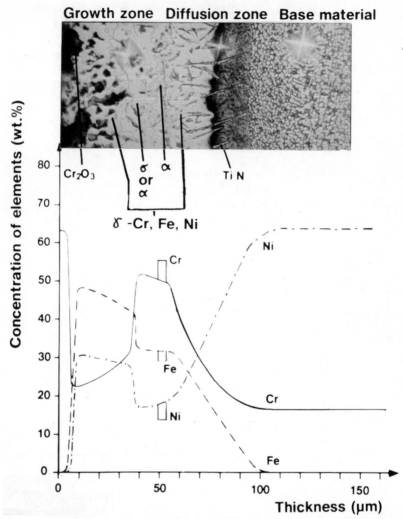

FIG. 5.   Chromium diffusion coating on IN-738 LC after operating test at ~850 °C

In this case, partly due to the more severe conditions of corrosion, the outer iron-rich γ-zone, of similar composition, overlies a relatively narrow middle σ-zone into which fingers or channels of the outer oxide have penetrated. The inner part of this zone contains a comparable volume of the α-chromium precipitate which becomes noticeably finer approaching the diffusion zone.

At higher temperatures, up to a maximum of 850 °C, no significant changes in the distribution of the respective phases are seen to occur (Fig. 5). As in the previous case there is a thick scale of Cr_2O_3 beneath which the metallic γ-phase is gradually depleted in chromium by the advancing oxide. The composition of the middle zone (51 % Cr, 32 % Fe and 17 % Ni) is similar to that formed at lower temperatures. The presence of the σ-phase could not be confirmed by X-ray analysis, so this composition may have corresponded to an α-phase.

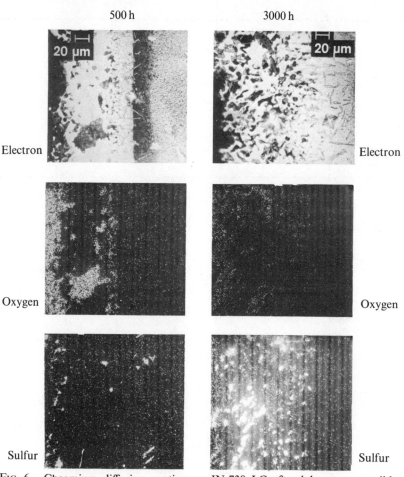

FIG. 6. Chromium diffusion coating on IN-738 LC after laboratory crucible testing at 850 °C

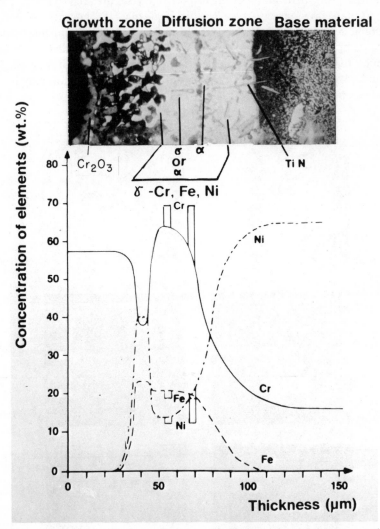

FIG. 7.   Chromium diffusion coating on IN-738 LC after laboratory crucible
testing at 850 °C/1000 h

CONDITION OF THE COATING AFTER LABORATORY
CRUCIBLE TESTING

In order to simulate the aggressive conditions of a gas turbine operating on blast furnace gas, specimens were corroded in a synthetic slag composed of 4.6% Na_2SO_4, 19.4% $CaSO_4$, 24.0% Fe_2O_3, 19.8% $ZnSO_4$, 11.1% K_2SO_4, 3.0% MgO, 7.0% Al_2O_3 and 11.1% SiO_2.

Even after relatively short periods of corrosion (Fig. 6, left) the coating is seen to exhibit stages of attack similar to the in-service examples. Nevertheless, in contrast, isolated sulfide particles could be observed penetrating as far as the diffusion zone. The process of corrosion has been considerably accelerated; thus after 3000 h the entire growth zone has been consumed (Fig. 6, right). Between the heterogeneously distributed sulfides and oxides are the remains of the γ-solid solution; the σ- or α-phases appear also to have been fully used up.

The distribution of elements after medium testing times (1000 h) also shows that beneath the chromium-depleted zone, middle zones rich in chromium have appeared with a phase distribution very similar to that found in the blades corroded in service (Fig. 7). It is here worth mentioning the comparably high concentration of nickel found in the outer γ-phase near the corrosion front. This possibly stems from the especially active sulfidation, a consequence of laboratory testing, which has yet to be investigated.

DISCUSSION

The examples demonstrate that in all cases a similar phase equilibrium exists within the scales regardless of the composition of the base material, up to a maximum temperature of 850 °C. It is therefore reasonable to expect that in all cases similar mechanisms of consumption occur. From the analysis of the phases observed a complete description of the processes is possible, using the NiCrFe phase diagram. Figure 8 shows the 650 and 800 °C isotherms of this system (9). Unfortunately, since no reliable information regarding the 850 °C isotherm is at present available the following discussion and conclusions must be based on the slightly lower temperature.

The diffusion coating as produced—somewhat influenced by the subsequent heat treatment of the base metal (8 h, 1080 °C; 24 h, 850 °C; 16 h, 730 °C)—is characterized by a chromium-rich α-phase (A), a heterogenous

380   *R. Bauer, H. W. Grünling and K. Schneider*

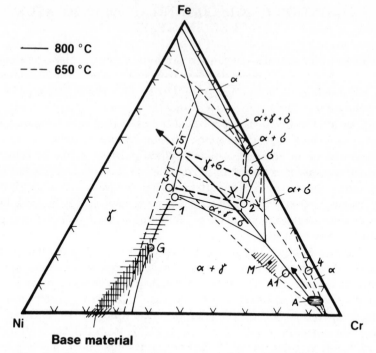

FIG. 8.   Phase diagram NiCrFe (from Pfeiffer and Thomas, ref. 9)

matrix of mean composition (A1) and, in equilibrium with this, a solid solution at the boundary to the base metal (G). Following the cross-hatched region (G, base metal) the diffusion zone reaches the composition of the base metal.

Based upon the composition of the matrix at the temperatures considered (710–730 °C), one would expect the appearance of σ-phase together with an iron-rich γ-solid solution and a chromium-depleted α-phase. The limiting compositions are fixed by the corner of the triangular α-γ-σ region of the ternary phase diagram.

The phases present after 35 000 h service may be similarly related. At the boundary to the base metal there exists an iron-rich γ-solid solution (1 in Fig. 8) in equilibrium with the σ-matrix(2) of the overlying zone, the inner portion of which contains γ(3) and α-Cr(4) precipitates. The composition of the phases may correspond to the corners of the three phases α-γ-σ triangle of the ternary phase diagram. Towards the outer part of the middle zone the α-phase disappears and the structure consists solely of σ(2) and γ(3). Within a thin boundary layer adjacent to the outer-lying γ-zone(5) the composition of the σ-phase has shifted towards a higher level of iron, $\sigma(2)$–$\sigma(6)$. Here also, an equilibrium relationship between the outer-lying $\gamma(5)$ and the adjoining $\sigma(6)$ can be recognized.

At the base metal side of the diffusion zone the composition of the iron-rich $\gamma(1)$ tends, according to the hatched line, towards that of the base metal.

The reactions within the coating are thus essentially determined by the consumption of chromium at the surface due to corrosion. The composition alters according to the Cr–X line with a fairly constant Fe/Ni content until at point X only $\sigma(2)$ and $\gamma(3)$ remain in equilibrium. This concentration shift corresponds, as one would expect, to a single phase base metal (in this case of Cr, Fe and Ni) in which one of the component elements (in this case Cr) is selectively oxidized (assuming a low O_2 solubility).

The relative concentrations of elements in the thereby chromium-depleted (Ni/Fe enriched) zone bordering the oxide are determined by the ratio of the oxidation rate-constant to the diffusion coefficient of chromium in the coating (chromium diffusion to the outside of the coating). The chromium content of the iron-rich γ-solid solution(5) at the surface of about 20% corresponds to that composition required to maintain the Cr_2O_3 protective coating, by diffusion of chromium back from the σ-zone. That this concentration adjusts equally to both high and low temperatures shows, on the basis of theoretical considerations, that the rate of oxidation is determined by the transport rate of chromium. This does not seem to be a process of homogeneous diffusion of chromium from the σ-zone along the dotted line X–5 but rather one of the dissolution of σ at the phase boundary. The σ-zone, then, seems to act as a 'chromium reservoir', whilst the outer γ-zone may be referred to as the 'working layer'. Thus, only when the σ-zone has been completely used up can uncontrolled corrosion attack set in, due to chromium depletion and the subsequent formation of mixed oxides.

From these conclusions, three stages may be formulated in the consumption of the coating:

(1) Equilibrium establishment and the consequent formation of a σ-zone as a result of the use of chromium for oxide formation at the surface; development of a γ-working layer(5).

(2) Controlled using-up of the coating through the consumption of the σ-zone, which acts as a chromium reservoir layer.

(3) Uncontrolled corrosion attack after the formation of mixed oxides due to chromium depletion in the coating.

At higher temperatures the existence of the σ-phase has not satisfactorily been established; nevertheless an α-phase of approximately similar composition could possibly fulfil the function of a chromium reservoir.

In comparable corrosive environments the accelerated consumption of the coating is based upon the temperature-dependent increase in the rate of diffusion of chromium out of the γ-working layer and the consequent more rapid consumption of the reservoir zone.

The laboratory tests indicate that as well as the initial composition of the coating and base metal and their operating temperature, the corrosive environment itself may also determine whether or not the process reaches stage 2. Due to the low rate of the diffusion processes and the sluggish adjustment to equilibrium within this system it may be contemplated that the corrosion of the coating will occur on a broad front, because the rate of corrosion is faster than that of the diffusion of chromium through the coating. This could explain the severe corrosion seen in Fig. 6. At the same time the work demonstrates the problem of applying accelerated laboratory corrosion tests under severe corrosive environments to the evaluation of protective coatings and alloys.

REFERENCES

1. H. W. Grünling (1974). 'High Temperature Corrosion Resistant Coatings in Stationary Gas Turbines', DGM Tech. Reps. of Mtg on Composite Materials, Constance.
2. G. Faber (1974). 'Protective Chromium Coatings for Stationary Gas Turbines', Symp. of European Corrosion Federation, Copenhagen.
3. W. Möller (1966). Progress in the fight against high temperature corrosion of gas turbine blades, *Brown Boveri Co. News*, **48**, 669–78.
4. M. Villat and P. Felise (1976). High temperature corrosion resistant coatings for gas turbines, *Techn. Rundschau, Sulzer* **3**, 97.
5. G. Lehnert and H. Meinhardt (1972). 'Developments in the Field of Metallic Diffusion Coatings for Protection Against High Temperature Corrosion', DGM Tech. Reps. of Mtg on Composite Materials, Constance.
6. P. R. Sahm (1976). High temperature corrosion and protective coatings, *Metall.*, **30**(4), 326–31.
7. H. W. Grünling (1975). 'High Temperature Corrosion in Gas Turbines', VDI Rep. No. 235, p. 155–62.
8. H. W. Grünling (1974). Operational experience with protective coatings in stationary gas turbines, lecture delivered at the Mtg on Techniques to Minimise Hot Corrosion by Coatings, Additives and Fuel Treatment, Copenhagen.
9. H. Pfeiffer and H. Thomas (1963). *Corrosion Resistant Alloys*, Springer, Berlin, p. 184–5.

DISCUSSION

W. Möller (BBC, Mannheim, Germany): (1) You spoke of a slag in which you performed the laboratory tests. Does the expression 'slag' mean that the mixture was totally or partly molten? Did you determine the melting point of the slag?

(2) Did you observe chromium sulfides in or beneath the coating after 35 000 h service at 710–730 °C?

Reply: (1) Thermal analysis has shown the slag to be an SO_2/SO_3 donor, forming no totally or partly molten phases up to a temperature of about 1000 °C.

(2) As we have said, after 35 000 h service, varying thicknesses of coating remained on the blade. In areas where much of the coating had been consumed, a few isolated chromium sulfides were seen, both in the 'oxide channels' and at the tip of the penetrating oxide. However, an uncoated neighboring blade of the same material showed a broad advancing front of chromium sulfide.

R. C. Hurst (Euratom, Petten, Netherlands): Why do you use large quantities of FeO in your laboratory slags, as it is now well known that iron is capable of catalysing the hot corrosion reaction producing, in some of my experiments, up to two orders of magnitude more corrosion in a wide range of Ni-base alloys?

Reply: The composition of the synthetic slag is based upon an average analysis of actual blade deposits over a period of years, and in particular iron oxide as hematite was included. To be more particular, its corrosive effect is to regulate the production of SO_3 formed by the decomposition of zinc sulfate and the subsequent formation of an iron zinc spinel.

E. Erdoes (Sulzer Bros. Ltd, Winterthur, Switzerland), additional reply: (a) Fe_2O_3 catalyses the SO_2/SO_3 equilibrium; (b) Fe_2O_3 in the slag reacts with ZnO (formed by $ZnSO_4 \rightarrow ZnO + SO_3$) to give $ZnFe_2O_4$.

D. B. Meadowcroft (CERL, Leatherhead, UK): There was a significant level of iron in the coating which was presumably added deliberately. Why was this?

Reply (by W. Möller, BBC, Mannheim, Germany): A comment on the question of D. B. Meadowcroft, concerning the origin of the iron in the coatings. During the commercial pack-cementation process the gas turbine blades are embedded in a powder of FeCr alloys. Therefore, Fe is transported via the gas phase in addition to the Cr. The result is a coating which contains $\sim 35\%$ Fe.

I. Ali-Khan (Kernforschungsanlage, Jülich, Germany): The surface of the Cr coatings you showed on IN-738 was very rough, meaning that the gas in contact with it could be stagnant. What is the effect of this roughness on the flow conditions in the gas?

The second question concerns the material's behavior without Cr coatings. Can you say something about this? Do you also have experience on the long-term behavior of Cr coatings, e.g. 10 years? Is a continuous reaction to be expected between coating and base material over a long time?

Reply: (1) Flaking of the outer oxide layers during specimen preparation gives the impression of a surface roughness higher than would occur during service. Nevertheless it is true that a greater roughness reduces the efficiency. To what extent is very much dependent on turbine blade design and on roughness developed in service by the oxide layers. In cooled blades especially, you would expect an increase in average material temperature because of changes in the heat transfer coefficient of the outer blade surface.

(2) Uncoated blades made of Nimonic 80A show far heavier corrosion attack than chromium-coated blades. There is hardly any change in chromium concentration gradients at the interface coating/base material up to 35 000 h service time. Therefore, we do not expect reaction of the coating with the base material, or even loss of chromium by inward diffusion.

P. Felix (BBC, Baden, Switzerland): First a remark on the comment of the previous speaker. Such Cr coatings have been proved in service up to 70 000 h

Now a question: Is the formation of the reservoir layer influenced by the original layer thickness? Can it be concluded from the fact that the formation of the reservoir layer is diffusion controlled, that at higher service temperatures substantially thicker layers are necessary so that the working layer/reservoir layer mechanism can operate?

Reply: The formation of the reservoir layer is not dependent upon the original coating thickness. There is a constant minimum thickness required for the production of the working layer and the diffusion zone, but above this the reservoir is always formed. Therefore, thicker chromium coatings will form thicker reservoir layers. At higher service temperature the diffusion-controlled processes will cause more rapid consumption of the coating; a thicker reservoir layer may partly compensate for this.

23

Preparation and Oxidation of Zirconium Silicide Coatings on Zirconium

M. Caillet, H. F. Ayedi, A. Galerie and J. Besson

Institut National Polytechnique de Grenoble, Saint Martin d'Hères, France

ABSTRACT

Coatings of $ZrSi/Zr_2Si$ on zirconium were prepared by deposition of silicon from monosilane on heated zirconium samples. The oxidation of these coatings by oxygen or water vapour follows the general rate law $(\Delta m)^n = kt$ ($n \simeq 1.6$). These kinetic results can be explained by the existence of two different silicide layers. At the lowest water vapour pressures, the kinetic behaviour is different; the formation of volatile SiO can account for this phenomenon.

The experiments have shown the good corrosion resistance of this coating compared with that of pure zirconium. For example, in oxygen atmospheres the weight gains are about eight times less for silicides than for zirconium.

INTRODUCTION

It is well known that metallic silicides have a good corrosion resistance at high temperatures and may therefore be used as protective coatings. That is the reason why, in our work on the oxidation of zirconium, after nitrides and carbonitrides (1), we prepared coatings of zirconium silicides and studied their oxidation behaviour. Atmospheres used were dry oxygen or water vapour.

Until now this problem has been tackled only by Kieffer *et al.* (2). In their work, zirconium silicides were oxidized in air to produce zirconia and amorphous silica.

SAMPLE PREPARATION

The coatings of zirconium silicides are prepared by deposition of silicon on an electrically heated tape of zirconium from a static atmosphere of monosilane (SiH_4). The possibility of the monosilane catching fire, due to an accidental air ingress, requires the use of a metallic apparatus. This is shown in Fig. 1. The zirconium sample is contained in a water-cooled reaction vessel; the thermal decomposition of SiH_4 occurs, therefore, only on the sample.

The temperature is measured through an optical quartz window by a single-colour pyrometer. A germanium photodiode, in front of the sample, gives an electrical signal which is amplified. It can be used to keep the initial brightness temperature of the sample constant (within $\mp 2\,°C$). The emissivity of silicon (0.64 for $\lambda = 0.65\,\mu m$) is, however, different from that of zirconium (0.35) and consequently the real temperature of the sample can be changed by the initial deposition of silicon; the temperature change never exceeds $40\,°C$.

At the end of the experiment, the reaction vessel can be purged with

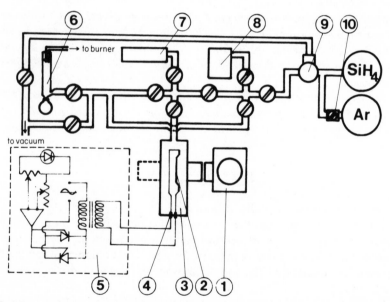

FIG. 1.   Apparatus for coatings preparation: 1, optical pyrometer; 2, metal strip; 3, reaction vessel; 4, sealed-in leads; 5, temperature control system; 6, gas outlet; 7, Pirani gauge; 8, ionization gauge; 9, manometer; 10, needle valve

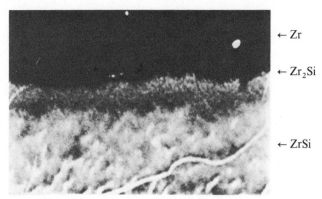

FIG. 2. Electronic image of a taper section of the zirconium (exact scale-markers are not given for Figs. 2, 3 and 4 because of the uncertainty in the taper angle)

flowing argon. The purging gas stream flows through the flame of a gas burner to oxidize the reaction hydrogen and the remaining monosilane.

The silicon deposit is formed on the zirconium tape heated to $\sim 1000\,°C$. The initial gas pressure is 5 torr. It is known that thermal decomposition of SiH_4 is rapid in these conditions. The limiting process is the monosilane diffusion in the gas phase (3–5). In a few minutes a layer of silicon can be obtained on the sample, but this layer is non-adherent. Therefore, the sample must be annealed at the same temperature for 1 h to permit the diffusion in the solid state. An adherent coating with a very bright metallic appearance is then obtained. X-ray analysis shows the presence in this coating of the two compounds $ZrSi$ and Zr_2Si (6, 7).

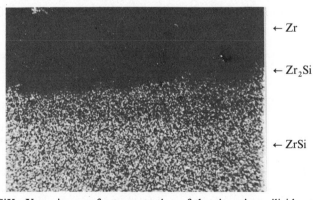

FIG. 3. $SiK\alpha$ X-ray image of a taper section of the zirconium silicides coating

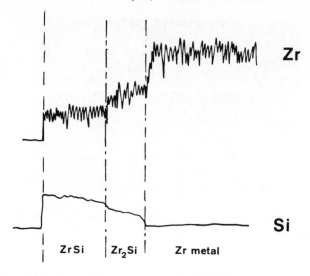

FIG. 4.   Concentration profiles of zirconium and silicon in the coating

The total thickness of the coating is 10 μm. Because of this low value, the microscopic examination of cross-sections of the samples did not reveal the presence of different layers. Sections at very low angles to the surface can magnify the apparent thickness of the coating. By this means Castaing's electron microprobe analysis technique can give useful results; Fig. 2 shows the electronic image of such a section. Two layers are clearly visible. The X-ray image of silicon (Fig. 3) and the profiles of the concentrations of silicon and zirconium (Fig. 4) corroborate the existence of these two layers: a layer of Zr_2Si inside, in contact with the metal and a layer of ZrSi on the outside.

OXIDATION OF THE COATING BY OXYGEN

Reaction Rate

The reaction rate was measured by isothermal thermogravimetry in a Cahn thermobalance within the range 700–900 °C and for oxygen pressures between 50 and 500 torr. In all cases the behaviour of the coated material is better than that of the pure zirconium. Some kinetic curves obtained with an oxygen pressure of 200 torr are shown in Fig. 5. The reaction rate is continuously decreasing. The equation $(\Delta m)^n = kt$ can describe the observed curves, as can be seen from the logarithmic plot $\log \Delta m = f(\log t)$

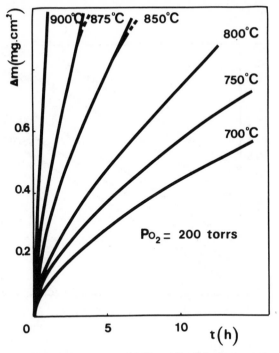

FIG. 5.   Oxidation by oxygen: isothermal weight increase vs. time

(Fig. 6). The value of n, calculated from this diagram, is approximately 1.63. The oxidation rate law is then

$$(\Delta m)^{1.63} = kt$$

The values of k can also be obtained at different temperatures and transferred to an Arrhenius plot. The resulting diagram (Fig. 7) shows that Arrhenius' law is not obeyed: the activation energy depends on the temperature.

Neither the value of n nor of k are influenced by oxygen pressure in the range 50–500 torr.

Reaction Products
X-ray examination of the scale after reaction indicates the presence of monoclinic ZrO_2 but SiO_2 is also thought to be present in an amorphous form: infrared spectrophotometry gives some spectra with a weak SiO_2 peak.

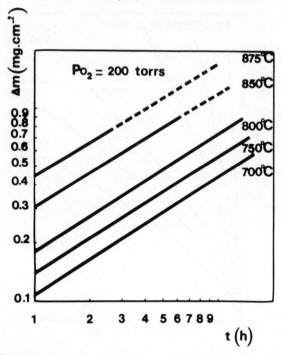

FIG. 6.   Oxidation by oxygen: log isothermal weight increase vs. log time

Discussion of the Results

The complexity of the ZrSi/Zr₂Si coating does not allow more than a qualitative explanation of the oxidation mechanism.

It can be supposed that the ratios Si:Zr in the oxide scale are the same as in the silicides. The oxidation products would then be $ZrO_2 + SiO_2$ and $2ZrO_2 + SiO_2$. These products are certainly compact, allowing for the fact that the Pilling–Bedworth ratios are, respectively,

$$f\left(\frac{ZrO_2 + SiO_2}{ZrSi}\right) = 2.23$$

$$f'\left(\frac{2ZrO_2 + SiO_2}{Zr_2Si}\right) = 1.95$$

Also the observed rate law corresponds approximately to a diffusion mechanism. Moreover, it can be seen in Fig. 4 that there is a silicon gradient in each layer of the initial coating.

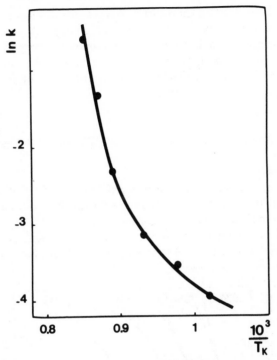

FIG. 7. Oxidation by oxygen: influence of temperature

Lastly, it is known that oxygen diffusion is faster in zirconia than in silica $(D_{ZrO_2}^{800°C} = 4.5 \times 10^{-14} \, cm^2/s; D_{SiO_2}^{800°C} = 3.7 \times 10^{-15} \, cm^2/s)$ (8).

Consequently it might be thought that the oxidation of the outer layer of ZrSi would begin according to a parabolic law. But it was shown that the diffusion coefficient of oxygen increases with the depth of penetration of this element; therefore, a positive deviation would be observed from the parabolic law. When the oxidant arrives at the layer of Zr_2Si, the mixed oxide formed is less protective than that formed at the beginning of the reaction: the deviation from the parabolic law increases. The rate law is then represented by a power law with an exponent between 1 and 2, as shown by the experiments.

The absence of an effect of oxygen pressure is difficult to elucidate. Only the following remarks can be made:

1. There is no influence of oxygen pressure on the oxidation rate of metals forming n-type semiconducting oxides (9) and zirconia is such an oxide (10).

2.   If the solid-state transport of oxygen in silica involves interstitial ions, it must be supposed that the amount of oxygen adsorbed corresponds to the asymptotic amount of the Langmuir isotherms.

3.   If the transport of oxygen in silica takes place through grain boundary short-circuit paths, then little effect of pressure on the rate law would be expected.

OXIDATION OF THE COATING BY WATER VAPOUR

Kinetic Results

The reaction rate was measured between 750 and 880 °C in the pressure range 0.3–15 torr. At 880 °C, the coating is completely oxidized after 10 min but is more resistant than the pure metal.

The thermogravimetric curves are shown in Fig. 8. A decrease of the reaction rate in the course of time may be noted, but the reaction rate is

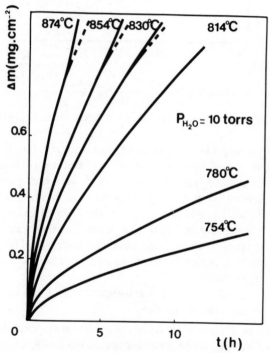

FIG. 8.   Oxidation by water vapour: isothermal weight increase vs. time

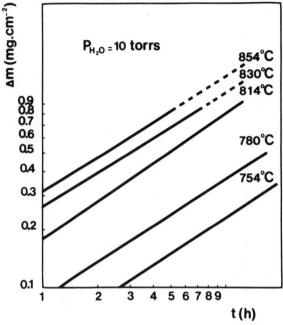

FIG. 9.   Oxidation by water vapour: log isothermal weight increase vs. log time

lower than in pure oxygen. In a logarithmic diagram, it can be seen that the weight gain of the sample is correctly represented by the equation: $(\Delta m)^n = kt$, the values of n being between 1.55 and 1.67 (Fig. 9).

The dependence of $\ln k$ upon $1/T$ is approximately linear (Fig. 10), so Arrhenius' law is obeyed and the value of the activation energy is about 66 kcal/mol.

The influence of the water vapour pressure between 0.3 and 15 torr, studied at 830 °C, is shown in Fig. 11. For the higher pressures ($p_{H_2O} \geq 1$ torr), the general shape of the curves is similar but it is impossible to give a mathematical expression for the pressure law $k = f(p_{H_2O})$. For the lower pressures ($p_{H_2O} \leq 1$ torr), the general law $(\Delta m)^n = kt$ no longer holds: the reaction rate increases with time. The pressure of 1 torr can thus be considered as a critical pressure for which a change in the mechanism occurs.

Reaction Products
As with oxidation in oxygen, monoclinic zirconia and amorphous silica have been identified in the oxide scale.

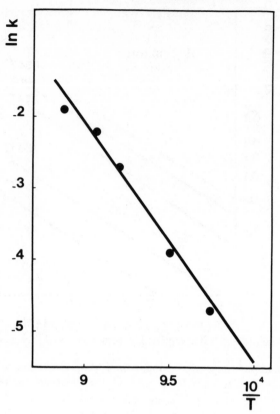

FIG. 10.   Oxidation by water vapour: influence of the temperature

DISCUSSION

Two cases are to be considered according to the water vapour pressures.

1. For pressures over 1 torr, the mechanism described for oxygen is still valid; the oxide scale is compact and the transport of oxygen involves atoms or ions resulting from the dissociative adsorption of the water vapour.

2. For pressures under 1 torr, the formation of volatile SiO must be considered; according to Wagner (11) this compound is stable at 830°C for oxygen pressures below about 10^{-5} atm. The oxygen partial pressure resulting from the thermal dissociation of water

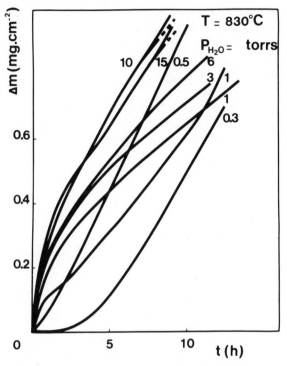

FIG. 11. Oxidation by water vapour: influence of the water vapour pressure

vapour ($T = 830\,°C$, $p_{H_2O} = 1$ torr) is 1.5×10^{-5} atm. Therefore, the formation of volatile SiO is thermodynamically possible for water vapour pressures below 1 torr. This SiO release results in the formation of cracks in the scale and the reaction rate becomes approximately constant as it is no longer diffusion controlled.

However, only a small proportion of the silicon must escape as SiO because the formation of zirconia alone (without any SiO_2) could not account for the weight gain of the sample.

CONCLUSIONS

1. Coatings of $ZrSi/Zr_2Si$ on zirconium were prepared by deposition of silicon from monosilane on heated zirconium samples.

2. The oxidation of these coatings by oxygen or water vapour followed the general rate law $(\Delta m)^n = kt$ $(n \simeq 1.6)$.
3. At the lowest water vapour pressures, the kinetic behaviour was different and can be accounted for by the formation of volatile SiO.
4. The detailed kinetic results can be explained by the existence of two different silicide layers.
5. The work has shown the good corrosion resistance of this coating compared with that of pure zirconium; in oxygen atmospheres the weight gains are about eight times less for the silicide coatings than for zirconium.

REFERENCES

1. H. F. Ayedi (1976). Thesis, Grenoble.
2. R. Kieffer, F. Benesovsky and R. Machenschalk (1954). Z. Metall., 45(8), 493.
3. F. C. Eversteyn, P. J. W. Severin, C. H. J. Vd Brekal and H. L. Peek (1970). J. Electrochem. Soc., 117, 925.
4. F. C. Eversteyn (1971). Philips Res. Repts, 26, 134.
5. J. Trilhe (1975). Thesis, Grenoble.
6. M. Hansen (1958). Constitution of Binary Alloys, McGraw-Hill, New York, p. 1206.
7. A. Madeyskii and W. W. Smeltzer (1969). Mat. Res. Bull., 3, 369.
8. B. E. Deal and A. S. Grove (1965). J. Appl. Phys., 36, 3770.
9. J. Besson and P. Sarrazin (1967). J. Chim., Phys., 64, 852.
10. D. L. Douglass (1971). The Metallurgy of Zirconium, Atomic Energy Review, Vienna.
11. C. Wagner (1952). J. Electrochem. Soc., 99, 369.

DISCUSSION

P. Felix (BBC, Baden, Switzerland): Does there exist a temperature where the oxidation rate of Zr is already significant but the oxidation rate of ZrSi coatings is still insignificant?

Reply: Yes, a temperature range exists for which the oxidation rate of zirconium is already important while that of the silicide coating is still negligible; this temperature range is about 500–600 °C.

Index

Acoustic emission technique, 175–83
  conclusions, 181
  experimental method, 177
  frequency spectrum analysis, 182
  results, 178–81
  single transducer, 177
  two transducers, 179
Al_2O_3 scale formation, 20, 61, 63, 64,
  65, 107, 113–17, 134
Alkali metal sulphate, 366
Alkaline earth sulphates, 166–9
Alumina
  inclusions, 274
  particles, 291
Aluminide coatings, 211, 249, 253–69
  coating thickness effect, 253
  dependence of diffusion on Cr
    content in Ni-base alloys,
    261–2
  dependence of layer structure on
    composition of base material,
    256–9
  diffusion barrier formation, 269
  effect on mechanical properties of
    substrate, 253–4
  influence of Ni content of substrate
    on layer structure, 259–61
  interlayer formation, 262–5
  iron-base alloys, on, 255
  layer formation, 262
  layer structure thickness, 267
  nickel-base alloys, on, 255–6
  see also Nickel aluminide coatings
Aluminium-base coatings, 201

Aluminium diffusion-type coatings,
  see Aluminide coatings
Aluminized coatings, *see* Chromium-
  aluminized coatings; Cerium
  chromium-aluminized coatings;
  Yttrium chromium-aluminized
  coatings
ATS 290-G, 258
ATS 340, 258
ATS 351, 258
Auger electron spectrum, 101

Barrier layers, 261–3
Boride coating, 214, 229
Burner-rig tests, 105–19, 121

Calcium, 156
  sulphate, 165
Capillary tube materials, 38
Ceramic bonding material with
  hydraulic cold-hardening, 152,
  156
Ceramics
  chemically bonded, 154–60
  mechanical properties, 174
  silicon-based, 121, 126, 131–3
Cerium chromium-aluminized
  coatings, 293–312
Chemical vapour deposition (CVD),
  196–7, 207, 211–12, 241–3, 314,
  333–51, 356
Chromaluminizing of IN-100, 283, 290

Chromesco III, 151
Chromic oxide scale, 226
Chromium
  austenitic alloys, 19
  boride, 214, 229
  coatings, 251
  diffusion coatings, 369–85
    as produced, 370–3, 379
    chromium transport mechanism, 370
    coating after crucible testing in synthetic slag, 379, 383
    condition after service, 373–7, 380
    consumption stages, 381–2
    diffusion zone, 370–1, 376, 379, 381
    growth zone, 370–1
    life expectancy, 370
    pack cementation, 384
    phase analysis, 379
    phase equilibrium, 379
    protective mechanism, 370
    reservoir layer, 384–5
    temperature effects, 375–7
    sulphides, 383
Chromium-aluminized coatings, 265–8, 293–312
  diffusion barrier formation, 269
  interlayer formation, 268
  layer structure and diffusion behaviour and base material composition, 267
  layer structure thickness, 267
Chronophotography, 148
Cobalt, silicide coated, 215
Co–Cr–Al alloys
  effect of Y and Hf additions on oxidation, 55–70
  microstructure, 57
CoCrAlY coatings, 247, 251
Combustion gases, oil-fired, 121–38
Co–Ni steel, 151
Corrosion–erosion
  phenomena, 139–60
  rates, 148
Corrosion rate, definition, 75
Corrosion rig and fuel composition data, 73

Cracking
  oxide layers, of, 175–83
  protective coatings, in, 203, 243, 248, 252
Creep
  behaviour of gas turbine materials, 87–103
  behaviour of IN-100, 191
  curves, 101
  deformation, 24
  rupture life, 207
  rupture tests of coated alloys, 196
  strength, 102, 103
  stresses, 203
  tests, 102
Cristobalite, 166
Cr–Ni alloys, 121–5, 129, 133–5, 138
Cr–Ni austenitic steels in helium with water vapour and hydrogen additions, 21–38
Cr–Ni steel, 151, 182
Cr_2O_3
  formation, 108–10
  growth rate, 49
  particles, 134
  scale, 43, 107, 133–5, 183, 220, 377
$Cr_{23}C_6$ layer, 225
Crucible tests, 166, 379
C73, silicide-coated nickel on, 222–6
C73/2A1, 243

Diffusion
  annealing, 279, 281
  barriers, 262–5, 268, 269, 330
  boundary layer, 34
  coefficient, 50, 346–8, 393
DP24, 247
Dragon project, 37
Ductile–brittle transition of high temperature coatings for turbine blades, 233–52
Ductile fracture, 94
Ductile grain boundary phase, 99–100

Elastic wave detection and analysis, 177

Electron microprobe analyses, 75, 340, 345, 373, 390
Enamelling, 152
Enstatite, 169

Failure mechanisms, 93
Fatigue
  high cycle, 209
  low cycle, 91, 209
  strength, 88
  stress, 203
Fatigue stress/time-to-rupture curves for coated and uncoated gas turbine blade materials, 193, 195
Fe–Cr–Al ferritic steels in sulphur, 1–20
Fe–Cr alloys, 121–5, 129, 133–5, 138
Fe–Cr–Ni alloys, 121–5, 127–9, 133–5
Forsterite, 169
Fracture characteristics, 94
Fuel element cladding materials, 21
Fuels, sulphur in, 102
Furnaces, corrosion–erosion and protection techniques in, 139–60

Galvanostatic measurements, 309–11
Gas turbine blade materials, fatigue stress/time-to-rupture curves for coated and uncoated, 193, 195
Gas turbine blades
  chemical loading, 188–9
  corrosion life, 201–2
  lifetime, 209–10
  masking, 211
  mechanical loading, 188
  mechanical properties, 206–7
  protective coatings for, *see* Protective coatings
  rupture strength, 196
  thermal loading, 188
Gas turbine materials, creep behaviour, 87–103
Gas turbines, silicon nitride in, 162

Grain boundary surfaces
  coatings, 94
  fracture, 99
Grain size effects in IN-738 alloy, 71–85
Gravimetric experiments, 27

Hafnium effect on oxidation behaviour of Co–Cr–Al alloys, 55–70
Hastelloy X, 258, 263, 308
Heat exchanger surfaces, 139
Heat treatment effects, 207, 243, 269, 295
Helium-cooled fast breeder reactor, 21
High vacuum techniques, 37
Hot-gas corrosion, 88
Hydrogen additions, Cr–Ni austenitic steels in helium with, 21
Hydrogen metering device, 26
Hydrogen sulphide/hydrogen mixtures, corrosion in, 3

Incineration of household refuse, 139–60
Incoloy-100, 103, 107, 113–18, 257, 258, 329, 330
  chromaluminizing, 283, 290
  creep behaviour, 191
  creep tests on, 87–103
  high activity type coatings, 283–6
Incoloy-800, 29, 151
Inconel-601, 136–7, 258, 265, 322
Inconel-713LC, 103, 117, 258, 302
Inconel-738, 71–85, 103, 107–10, 118, 258, 347, 384
Inconel-738 LC, 203, 207, 217–20, 229–30, 243, 245, 247, 301, 303, 307, 309, 375
Inconel-935, 257, 258
Inconel-939, 243, 245, 247
Intergranular corrosion, 34
Intergranular fracture, 94
Intergranular oxidation, 35–7, 50
Intermetallic particle distribution, 50

Internal degradation depths, 49
Interstitial impurities, 49
Ion bombardment, 350
Iron, 173
Iron-base alloys, 255, 268
Issy-les-Moulineaux plant, 141
Ivry-sur-Seine plant, 141

Kerosene containing impurities,
    combustion products, of,
    105–19
Kirkendall
    effect, 314
    holes, 219
    porosities, 273, 277, 280

Layer recrystallization, 16
Lifetime limits, 187–8, 194, 196

Magnesium, 169, 173
    compounds, injection of, 152
    sulphate, 165
MAR-M002, 103, 258, 329
MAR-M246, 103, 117, 258
Mechanical properties, changes in, 35,
    36
Meinhardt, 255
Metallization techniques, 152
Metallographic investigations, 29–33,
    41
Metallographic structure, 286
MgO additions, 137, 174
Microanalysis, 33–5
Microfractographic examinations, 93,
    99–100
Micrographic observations, 311
Microhardness measurements, 34–5
Microstructure examination, 75, 92,
    243, 295
Molybdenum, 263
Morphological examination, 47, 94,
    109, 117, 128, 134

NiAl, 264, 269, 325

Ni_2Al_3 layer formation, 279, 280
Nickel, 347
    C73, on, 222–6
    NiAl-type coatings, 273–7, 280
    silicide-coated, 214
Nickel aluminide coatings, 271–9
    alumina particles in, 291
    appropriate alloy deposit prior to
        low or high activity process,
        286–90
    direct formation processes, 271–9
    examples, 281–90
    formation of Ni_2Al_3 layer followed
        by high temperature diffusion
        annealing, 272, 279–81
    formation modes, 271–2
    high activity, 281, 283–6
    nickel, on, 273–7, 280
    nickel-base alloys, on, 277–80
Nickel aluminides, 256, 259–65
Nickel-base alloys, 271
    aluminide coatings on, 255–6
    barrier effect of CrNi interlayer,
        261
    chromium-aluminized, 268
    diffusion barriers, 268
    NiAl-type coatings, 277–80
    protective coatings on, *see*
        Protective coatings
Nickel silicides on IN-738LC, 217–20
Ni–Cr alloys, 258
NiCrAlY coatings, 247–8, 251
NiCrFe phase diagram, 379
Ni–Cr–Si coatings, 313–31
Ni–Cr–Si plasma coatings, 203
Ni–Cr–Si powder coatings, 317–8
Ni–Cr–Si system, 317
NiCrSiB coating on IN-738 LC,
    229–30
Ni–Cr–Zr alloys
    oxidation rate of, 398
    oxygen/sulphur dioxide
        atmosphere, in, 39–54
    zirconium silicide coatings on,
        387–98
Nimonic-80A, 242, 243, 247, 371,
    373, 384
Nimonic-90, 103, 319, 322, 324–6

Nimonic-105, 209, 211, 243, 245, 247, 343, 345, 347
Niobium, 263
Nitrogen embrittlement, 239
Notch effect, 195

Oil-fired combustion gases, 121–38
Oil-fired combustion rig tests, 171
Ostwald ripening, 229
Osumilite, 173
Oxidation
    behaviour, 52, 286
    Co–Cr–Al alloys, of, 55–70
    effects, 24
    intergranular, 22
    kinetics, isothermal, 57, 59
    oxide layer cracking during, 175–83
    reaction kinetics, 36
    resistance, 37, 72
    silicon nitride, of, 165
    tests, isothermal, 56
Oxide
    composition, 53
    dispersions, 72
    layers, cracking of, 175–83
    morphology, 60
    protrusions, 65
    whiskers, 63
Oxygen/sulphur dioxide atmospheres
    Ni–Cr–Zr alloys in, 39–54
    protective action of, 166

Parabolic rate constants, 43
Pegs formation, 64, 68
Penetration thickness as function of time, 75
Phase stability of high temperature coatings on NiCr-base alloys, 213–31
Phosphoric acid attack, 160
Pilling–Bedworth ratios, 392
Platinum, 263
    aluminides, 263
Potassium, 156
Potentiodynamic experiments, 307
Potentiostatic measurements, 308

Powder metallurgy techniques, 84–5
Protective coatings, 185–97
    adhesion, 206
    anisotropy, 234
    basic requirements, 190–1
    brittle mode elongation, 247
    chemical vapour deposition (CVD), 196–7, 207, 211–13, 241–3, 314
    corrosion resistance, 201–2
    cost considerations, 185–6
    cracking in, 203, 243, 248, 252
    ductile–brittle transition temperature, 233–52, 241–2
        brittle mode elongation, and, 249
        model requirements, 240
        parameters which can influence, 237–40
        theoretical considerations, 236–7
    economic factors, 209–10
    erosion resistance, 202–3
    gas turbine blades, 199–212
    glow discharge, 242
    influence on base-alloy mechanical properties, 206–9
    influence on blade functions, 209
    laboratory test facilities and results, 191–5
    lifetime limits, 187–8, 196
    lifetime of coated components, 194
    local defect simulation, 203
    mechanical properties, 234, 235
    mechanical strength, 203–6
    multilayer, 206
    nickel-base alloys, 293–312
    phase stability on NiCr base alloys, 213–31
    plasma sprayed, 214, 315
        carbides in, 210
        NiCrSi, 207
    reparability, 212
    requirements, 199–200, 233
    salt bath deposition, 242
    self-healing, 212
    service loadings, 187–9
    silicon-containing, *see* Silicon-containing coatings
    simulation of service conditions, 192

Protective coatings—*contd.*
  slurry process, 256, 295, 315
  strain to fracture, 241
  stress, and, 186
  thermal stability, 203
  usefulness of, 186
  wear resistance, 197
  see also under specific coating types

Radiantly heated walls, corrosion of, 141
Rare-earth effect, 56
Rare-earth metals, 72, 311, 312
Rare-earth oxides, 72
Rate constants, parabolic, 43
Reactant gas pressure effect, 16
Remanit 4301, 255
René-41, 87–103, 263

Salt-coated tests, 105–19
Scale
  adherence, 56, 60, 69
  formation, 50, 52, 53, 91
  growth, stages of, 14–17
  layers, 92
Scale/alloy interface, 54
Selected area diffraction (SAD) examination, 301
Silicide-coated cobalt, 215
Silicide-coated nickel, 214, 222–6
Silicide coatings, 314, 317
Silicide layers, 231
Silicon-based ceramics, 121, 126, 131–3
Silicon carbide, 125, 126
Silicon-containing coatings, 121, 126, 131–3, 213, 241, 243, 247, 313–31
  brittle fracture behaviour, 330–1
  chemical vapour deposition (CVD), 333–51, 356
  chloride compounds used, 342
  coating techniques, 354–7
  composition changes, 318
  corrosion rate, 365
  corrosion tests, 351, 358

Silicon-containing coatings—*contd.*
  critical temperature, 342
  diffusion behaviour, 328
  diffusion calculations, 339–48
  diffusion layers, 314
  effects of substrate material, 343
  experimental arrangement, 338
  experimental results, 339–48
  extent of corrosion under vanadic attack, 359–65
  gas phase composition, 351
  high silicon contents and additional reaction barriers, with 324–6
  ion plating, 357, 364
  kinetics, of, 338, 350
  kinetics of vanadic corrosion, 358
  mass deposition rate, 339
  morphological features, 365
  overlay, 314
  oxidation behaviour, 326–8
  pack coatings, 356
  plasma-sprayed, 357
  preparation methods, 314–17
  procedures for obtaining, 314
  reaction-sintered layers, 324–5
  reaction-sintered solid solution layers with low silicon content, 321–4
  reaction-sintering of powder coatings, 319–21
  room temperature flexural strength of whole composite, 328, 330
  solid-state diffusion, 343
  temperature effects, 360
  thermodynamic analysis, 334–8
  vanadic attack on, 353–67
  see also Zirconium silicide coatings
Silicon-enriched layers, *see* Silicon-containing coatings
Silicon nitride, 125, 126, 132
  corrosion behaviour in molten salts, 161–74
  crystallization, 174
  gas turbines, in, 162
  grades available, 162
  hot-pressed, 162
  mechanical behaviour, 174
  oxidation modes, 165

Silicon nitride—*contd.*
  reaction bonded, 162
  sulphur in corrosion of, 165
Sodium
  chloride, 105–19, 165
  naphthenates, 122
  silicate, 132
  sulphate, 105–19, 125, 131, 135, 165
  vanadate, 135
Sodium-cooled fast breeder reactor,
    22
Solidification structures, 340
Spallation, 56, 60, 66, 110
Sputtering effect, 351
Statistical study, 148–51
Strain rate, 237
Stress
  effects, 69, 91, 186, 193–6, 233
  relief, 175
Stress/strain curve, 243, 248
Sulphidation, internal, 20
Sulphide formation, 17, 54
Sulphide scale morphology, 2, 3
Sulphur, 156
  corrosion of silicon nitride, in, 165
  Fe–Cr–Al ferritic steels in, 1–20
  fuels, in, 102
  migration, 54
  vapour, corrosion in, 2
Superheaters, corrosion of, 143,
    148–51
Surface layer formation, 22
Surface roughness effects, 18–19, 189,
    195, 384

Temperature effects, 16, 150
Thermal analysis, 383
Thermal cycling tests, 60, 66
Thermal expansion coefficients, 206
Thermal shock, 329
  tests, 298
Thermal stresses, 177
Thermax 4841, 255
Thermax 4876, 255, 256
Thermobalance experiments, 105–19
Thermobalance measurement
  technique, 303

Thermodynamic analysis, 334–38
Thermodynamic considerations, 24
Thermodynamic data, 49
Thermodynamic measurements, 36,
    390
Thermogravimetric measurements, 394
Titanium
  carbides, 133–5, 286
  scale formation, 178
Tridymite, 166
Turbine blade hot corrosion, 105

Udimet 520, 102
Ultrasonic thickness measurements,
    148

Vacancy condensation, 69
Vanadium, 131, 135, 137
  naphthenates, 122
  pentoxide, 125, 126, 166, 173
  salts, 131
Vanadium-rich deposits attack on
    silicon-containing coatings,
    353–67
Void formation, 61, 65, 66

Water-injection method, 26
Water vapour additions, Cr–Ni
    austenitic steels in helium with,
    21
Water vapour, oxidation by, 394
Waveguides, stainless steel, 182
Weight increase, parabolic time
    dependence of, 36

X-ray analyses, 154, 157–9
X-ray diffraction, 154, 165, 169, 173,
    373

Y_2O_3, 71–85, 301, 311
Yttrium chromium-aluminized
    coatings, 293–312

Yttrium effect on oxidation behaviour
    of Co–Cr–Al alloys, 55–70

Zirconium silicide coatings, 387–98
  kinetics of, 390, 394
  oxidation by oxygen, 390–4
  oxidation by water vapour, 394–8

Zirconium silicide coatings—*contd.*
  oxidation mechanism, 392
  oxygen pressure, absence of, 393–4
  reaction products, 391, 395
  reaction rate, 390, 394
  sample preparation, 388–90
  temperature effects, 398
  thickness, 390